Bioactive Dental Materials

Bioactive Dental Materials

Guest Editors

Tobias T. Tauböck
Matej Par

Basel • Beijing • Wuhan • Barcelona • Belgrade • Novi Sad • Cluj • Manchester

Guest Editors

Tobias T. Tauböck
University of Zurich
Zurich
Switzerland

Matej Par
University of Zagreb
Zagreb
Croatia

Editorial Office
MDPI AG
Grosspeteranlage 5
4052 Basel, Switzerland

This is a reprint of the Special Issue, published open access by the journal *Materials* (ISSN 1996-1944), freely accessible at: https://www.mdpi.com/journal/materials/special_issues/dental_biomaterial.

For citation purposes, cite each article independently as indicated on the article page online and as indicated below:

Lastname, A.A.; Lastname, B.B. Article Title. *Journal Name* **Year**, *Volume Number*, Page Range.

ISBN 978-3-7258-2787-9 (Hbk)
ISBN 978-3-7258-2788-6 (PDF)
https://doi.org/10.3390/books978-3-7258-2788-6

Cover image courtesy of Matej Par

© 2024 by the authors. Articles in this book are Open Access and distributed under the Creative Commons Attribution (CC BY) license. The book as a whole is distributed by MDPI under the terms and conditions of the Creative Commons Attribution-NonCommercial-NoDerivs (CC BY-NC-ND) license (https://creativecommons.org/licenses/by-nc-nd/4.0/).

Contents

Danijela Marovic, Håvard J. Haugen, Visnja Negovetic Mandic, Matej Par, Kai Zheng, Zrinka Tarle and Aldo R. Boccaccini
Incorporation of Copper-Doped Mesoporous Bioactive Glass Nanospheres in Experimental Dental Composites: Chemical and Mechanical Characterization
Reprinted from: *Materials* **2021**, *14*, 2611, https://doi.org/10.3390/ma14102611 1

Camila Corral Nunez, Diego Altamirano Gaete, Miguel Maureira, Javier Martin and Cristian Covarrubias
Nanoparticles of Bioactive Glass Enhance Biodentine Bioactivity on Dental Pulp Stem Cells
Reprinted from: *Materials* **2021**, *14*, 2684, https://doi.org/10.3390/ma14102684 13

Ivan Salinovic, Zdravko Schauperl, Marijan Marcius and Ivana Miletic
The Effects of Three Remineralizing Agents on the Microhardness and Chemical Composition of Demineralized Enamel
Reprinted from: *Materials* **2021**, *14*, 6051, https://doi.org/10.3390/ma14206051 25

Darko Božić, Ivan Ćatović, Ana Badovinac, Larisa Musić, Matej Par and Anton Sculean
Treatment of Intrabony Defects with a Combination of Hyaluronic Acid and Deproteinized Porcine Bone Mineral
Reprinted from: *Materials* **2021**, *14*, 6795, https://doi.org/10.3390/ma14226795 34

Hung-Yang Lin, Yi-Jung Lu, Hsin-Hua Chou, Keng-Liang Ou, Bai-Hung Huang, Wen-Chien Lan, et al.
Biomimetic Ceramic Composite: Characterization, Cell Response, and In Vivo Biocompatibility
Reprinted from: *Materials* **2021**, *14*, 7374, https://doi.org/10.3390/ma14237374 46

Danijela Marovic, Matej Par, Karlo Posavec, Ivana Marić, Dominik Štajdohar, Alen Muradbegović, et al.
Long-Term Assessment of Contemporary Ion-Releasing Restorative Dental Materials
Reprinted from: *Materials* **2022**, *15*, 4042, https://doi.org/10.3390/ma15124042 59

David Donnermeyer, Patrick Schemkämper, Sebastian Bürklein and Edgar Schäfer
Short and Long-Term Solubility, Alkalizing Effect, and Thermal Persistence of Premixed Calcium Silicate-Based Sealers: AH Plus Bioceramic Sealer vs. Total Fill BC Sealer
Reprinted from: *Materials* **2022**, *15*, 7320, https://doi.org/10.3390/ma15207320 74

David Opšivač, Larisa Musić, Ana Badovinac, Anđelina Šekelja and Darko Božić
Therapeutic Manuka Honey as an Adjunct to Non-Surgical Periodontal Therapy: A 12-Month Follow-Up, Split-Mouth Pilot Study
Reprinted from: *Materials* **2023**, *16*, 1248, https://doi.org/10.3390/ma16031248 84

Ivan Šalinović, Falk Schwendicke, Haitham Askar, Jamila Yassine and Ivana Miletić
Effects of Ion-Releasing Materials on Dentine: Analysis of Microhardness, Appearance, and Chemical Composition
Reprinted from: *Materials* **2023**, *16*, 7310, https://doi.org/10.3390/ma16237310 94

Takaharu Abe, Ryo Kunimatsu and Kotaro Tanimoto
Comparison of Orthodontic Tooth Movement of Regenerated Bone Induced by Carbonated Hydroxyapatite or Deproteinized Bovine Bone Mineral in Beagle Dogs
Reprinted from: *Materials* **2024**, *17*, 112, https://doi.org/10.3390/ma17010112 106

Mario Ceddia, Luciano Lamberti and Bartolomeo Trentadue
FEA Comparison of the Mechanical Behavior of Three Dental Crown Materials: Enamel, Ceramic, and Zirconia
Reprinted from: *Materials* **2024**, *17*, 673, https://doi.org/10.3390/ma17030673 . . . . **116**

Iván Valdivia-Gandur, María Cristina Manzanares-Céspedes, Wilson Astudillo-Rozas, Oscar Aceituno-Antezana, Victòria Tallón-Walton and Víctor Beltrán
Simultaneous Osseo- and Odontointegration of Titanium Implants: Description of Two Cases in Human and Animal Models and Review of Their Experimental and Clinical Implications
Reprinted from: *Materials* **2024**, *17*, 5555, https://doi.org/10.3390/ma17225555 . . . . **135**

Article

Incorporation of Copper-Doped Mesoporous Bioactive Glass Nanospheres in Experimental Dental Composites: Chemical and Mechanical Characterization

Danijela Marovic [1,*], Håvard J. Haugen [2], Visnja Negovetic Mandic [1], Matej Par [1], Kai Zheng [3], Zrinka Tarle [1] and Aldo R. Boccaccini [3]

[1] Department of Endodontics and Restorative Dentistry, School of Dental Medicine, University of Zagreb, 10000 Zagreb, Croatia; vnegovetic@sfzg.hr (V.N.M.); mpar@sfzg.hr (M.P.); tarle@sfzg.hr (Z.T.)
[2] Department of Biomaterials, Institute of Clinical Dentistry, University of Oslo, 0455 Oslo, Norway; h.j.haugen@odont.uio.no
[3] Department of Materials Science and Engineering, Institute of Biomaterials, University of Erlangen-Nuremberg, 91058 Erlangen, Germany; kai.zheng@fau.de (K.Z.); aldo.boccaccini@fau.de (A.R.B.)
* Correspondence: marovic@sfzg.hr

Abstract: Experimental dental resin composites incorporating copper-doped mesoporous bioactive glass nanospheres (Cu-MBGN) were designed to impart antibacterial and remineralizing properties. The study evaluated the influence of Cu-MBGN on the mechanical properties and photopolymerization of resin composites. Cu-MBGN were synthesized using a microemulsion-assisted sol–gel method. Increasing amounts of Cu-MBGN (0, 1, 5, and 10 wt %) were added to the organic polymer matrix with inert glass micro- and nanofillers while maintaining a constant resin/filler ratio. Six tests were performed: X-ray diffraction, scanning electron microscopy, flexural strength (FS), flexural modulus (FM), Vickers microhardness (MH), and degree of conversion (DC). FS and MH of Cu-MBGN composites with silica fillers showed no deterioration with aging, with statistically similar results at 1 and 28 days. FM was not influenced by the addition of Cu-MBGN but was reduced for all tested materials after 28 days. The specimens with 1 and 5% Cu-MBGN had the highest FS, FM, MH, and DC values at 28 days, while controls with 45S5 bioactive glass had the lowest FM, FS, and MH. DC was high for all materials (83.7–93.0%). Cu-MBGN composites with silica have a potential for clinical implementation due to high DC and good mechanical properties with adequate resistance to aging.

Keywords: dental; resin composites; copper; bioactive glass; mesoporous; nanoparticles

1. Introduction

Conventional restorative materials have several drawbacks, with one of the greatest being the occurrence of caries surrounding them, the so-called secondary caries [1]. Secondary caries is the major cause for restoration replacement and causes a significant workload escalation and economic burden [2].

The issue of secondary caries has enticed our research group to develop remineralizing bioactive materials with the addition of amorphous calcium phosphate (ACP) [3–6] and 45S5 bioactive glass (BG) fillers [7–11]. The supersaturated calcium and phosphate ion concentrations released by ACP resin composites have been shown to induce hydroxyapatite formation and remineralization [12]. Silica nanofillers (SiO_2) have been shown to enhance apatite formation on the surface of both ACP and BG composites [4]. However, the amorphous structure and solubility of ACP fail to provide strong mechanical properties of resin composites [3–5]. Similar behavior was demonstrated by BG resin composites, whose mechanical properties and degree of conversion (DC) deteriorated in a dose-dependent manner with the increase in BG ratio [8,10]. None of the investigated bioactive fillers, i.e., neither ACP nor BG, were silanized.

In commercial composites, the silane coupling layer is a bifunctional molecule that simultaneously creates a covalent Si–O–Si bond to the filler particles and copolymerizes with the methacrylate groups in the resin phase. The synergy between the filler and the resin increases the mechanical strength of the composite material by arresting crack propagation [13]. On the other hand, silanization hinders the ion release from bioactive fillers and is commonly considered undesirable for maintaining the required filler reactivity [14–16].

To counteract the abovementioned issue and avoid silanization of bioactive fillers, mesoporous bioactive glass nanospheres seem to be a suitable alternative. Mesoporous particles have a higher surface area than dense spherical ones due to a porous structure. Theoretically, low-viscosity resin monomers could penetrate pores and remain mechanically interlocked in their position after polymerization [17]. Therefore, the necessity of chemical bonding through silanization could be circumvented because of the micromechanical interlocking of resin and bioactive mesoporous particles [18]. Besides, greater reactive surface of mesoporous bioactive particles provides enhanced bioactivity compared to conventional bioactive glass [19]. This fact reduces the need for high amounts of bioactive glass in resin composites to achieve the remineralizing effect.

Mesoporous bioactive glass is currently in the spotlight of regenerative medicine. In addition to dense bioactive glass particles possessing the ability to release calcium, phosphorus, and silicon ions [20], mesoporous particles can act as carriers of therapeutic ions, such as copper, silver, or zinc [21–25].

Zheng et al. developed copper-doped mesoporous bioactive nanospheres (Cu-MBGN) by a sol–gel method using a Cu/L–ascorbic acid complex as the precursor of Cu [24]. This method enabled better particle dispersion, which represents a common problem of nano-sized particles [26]. Synthesized spherical particles were 100–300 nm with 2–10 nm pore size [24]. Cu-MBGN are primarily synthesized for biomedical applications, particularly osteogenic regeneration [24]. Application of Cu-MBGN caused a reduction in bacterial viability of *Staphylococcus aureus*, *Escherichia coli*, and *Staphylococcus epidermidis* [19,27,28] as well as hydroxyapatite formation upon exposure to simulated body fluid [19,24].

If added to conventional resin monomers and silanized inert micro- and nanofillers, the end product could be a multifunctional dental composite with highly desirable properties, namely antibacterial, ion-releasing/remineralizing, and improved mechanical strength. To our knowledge, no similar composite materials have been synthesized and investigated so far.

This study aimed to examine the influence of the addition of Cu-MBGN with proven antibacterial effect on selected mechanical and curing properties of experimental resin composites. Two strategies were tested in this study:

1. Bimodal approach: A material containing only Cu-MBGN fillers and inert silanized microfillers was compared to control materials. For the inert control material, a composite containing inert silica nanofillers and inert microfillers was used. The bioactive control consisted of conventional bioactive glass 45S5 and inert microfillers.
2. Trimodal approach: Similar to commercial materials, three types of fillers were used, namely inert silanized microfillers, inert silanized silica nanofillers, and various amounts of unsilanized Cu-MBGN fillers. Inert and bioactive control materials had identical composition as in the bimodal approach but with the total filler amount adjusted to 70 wt %.

The first null hypothesis of the bimodal strategy was that the addition of Cu-MBGN would not influence the flexural strength (FS), flexural modulus (FM), Vickers microhardness (MH), and DC in comparison to inert silica fillers or commercial bioactive fillers. The second null hypothesis of the trimodal strategy was that the combination of silica and increasing amounts of Cu-MBGN would not improve the tested properties (FS, FM, MH, and DC) over Cu-MBGN alone.

2. Materials and Methods

2.1. Synthesis of Cu-MBGN

Cu-MBGN were synthesized using a microemulsion-assisted sol–gel method as reported in our previous work [24]; the detailed synthesis procedures can be found in the literature. The chemical composition of Cu-MBGN was 85.4SiO_2–10.2CaO–4.4CuO (in mol %) as reported in the literature [24].

2.2. X-ray Diffraction (XRD)

XRD patterns of Cu-MBGN and 45S5 BG (Schott AG, Mainz, Germany) powders were recorded at room temperature with a Panalytical Aeris powder diffractometer (Malvern Panalytical, Malvern, UK) using CuK$\alpha_{1,2}$ radiation.

2.3. Scanning Electron Microscope (SEM)

The commercial 45S5 BG and Cu-MBGN powders were sputter-coated with 5 nm Au/Pd layer (90% Au/10% Pd), evaporated with argon plasma at 6 kV using the Precision Etching Coating System (Model 682, Gatan Inc., Pleasanton, CA, USA).

FE-SEM images of prepared samples were taken with a scanning electron microscope JSM7000F (JEOL Ltd., Tokyo, Japan) linked to the energy-dispersive X-ray analyzer EDS/INCA 350 (Oxford Instrument, Abingdon, UK).

2.4. Mixing of Experimental Resin Composites

Bisphenol A–glycidyl methacrylate (BisGMA) and triethylene glycol dimethacrylate (TEGDMA) were purchased from Merck, Darmstadt, Germany. An identical resin matrix containing 60:40 weight ratio of BisGMA: TEGDMA with a photoinitiator system (camphorquinone (0.2 wt %; Merck) and ethyl-4-dimethylamino benzoate (0.8 wt %; Merck)) was used for all materials. Resin was heated to 60 °C prior to admixture of fillers.

Four types of fillers were used in this study, as shown in Table 1.

Table 1. Characteristics of fillers used in the present study. Data provided by the manufacturers.

Name	Type	Manufacturer/ Product	Composition (wt %)	Size	Silanization
Cu-MBGN	Experimental/ Bioactive	Laboratory made	SiO_2 84.8% CaO 9.4% CuO 5.8% *	~100 nm	No
45S5 bioactive glass	Commercial/ Bioactive	Schott, Germany G018-144	SiO_2 45% Na_2O 24.5% CaO 24.5% P_2O_5 6%	4.0 μm	No
Ba glass	Commercial/ Inert	Schott, Germany GM27884	SiO_2 55.0% BaO 25.0% B_2O_3 10.0% Al_2O_3 10.0%	1.0 μm	Yes 3.2%
Silica	Commercial/ Inert	Evonik Degussa, Germany Aerosil DT	SiO_2 > 99.8%	12 nm	Yes 4–6%

* composition determined by ICP-AES analysis, data from [24].

The materials were mixed in the absence of blue light using an asymmetrical centrifugal mixer (Speed Mixer TM DAC 150 FVZ, Hauschild & Co KG, Hamm, Germany) at gradually increasing speed up to 2700 rpm.

Eight experimental resin composites were prepared, divided into two groups:
1. Group testing the bimodal approach with 65 wt % total filler load and
2. Group testing the trimodal approach with 70 wt % total filler load used for investigating 1, 5, and 10 wt % Cu-MBGN composites with silica fillers.

Each group contained both inert control (silica and microfillers; 10-Si and 15-Si) and bioactive control (45S5 BG and microfillers, 10-BG and 15-BG) in adequate amounts corresponding to the total filler load (Table 2).

Table 2. Composition of experimental resin composites (all amounts in wt %).

Group	Material	Resin	Inert Microfillers	Silica Nanofillers	Cu-MBGN	45S5 BG
Bimodal approach (65% filler load)	10-CuBG	35%	55%	-	10%	-
	10-BG			-	-	10%
	10-Si			10%	-	-
Trimodal approach (70% filler load)	1-CuBG-Si	30%	55%	14%	1%	-
	5-CuBG-Si			10%	5%	-
	10-CuBG-Si			5%	10%	-
	15-BG			-	-	15%
	15-Si			15%	-	-

2.5. Flexural Strength and Modulus

FS and FM were measured according to ISO/DIN 4049:1998 using the three-point bending test [29]. Custom-made stainless steel mold with an opening and dimension of $16 \times 2 \times 2$ mm^3 was filled with the composite paste, covered with polyester strips on both sides to prevent oxygen-inhibited polymerization layer formation, and pressed with a weight to extrude excess material. The specimens were light-cured three times on each side with overlapping exposures, i.e., six times in total. The light guide of the curing unit was always in direct contact with the polyester strip. Composite specimens were photopolymerized for a total of 120 s (6×20 s) with a light-curing unit (Bluephase PowerCure, Ivoclar Vivadent, Schaan, Liechtenstein, 950 mW/cm^2). The upper side that was polymerized first was marked, and the specimens were stored in saline solution in the dark in an incubator at 37 ± 1 °C for 1 or 28 days. Ten specimens were prepared per material and time point.

After the designated time, the specimens were removed from the saline solution, dried, and tested immediately with the side that was irradiated first facing the jig. The customized universal testing machine Ultratester (Ultradent, Salt Lake City, UT, USA) was used at a 1 mm/min crosshead speed until specimen failure.

2.6. Vickers Microhardness

Vickers microhardness (MH) was measured using Vickers hardness testing machine CSV-10 (ESI Prüftechnik GmbH, Wendlingen, Germany). The specimens prepared for the three-point bending test and aged for 1 or 28 days in saline solution at 37 °C were polished to remove the surface resin-rich layer with 4000 grit silicon carbide paper and 0.05 µm aluminum oxide slurry. The measurements were made with a 100 g load and 15 s dwell time at the specimen surface. Six specimens per material were subjected to five measurements per depth. The data from different measuring points on the same specimen were pooled and treated as a statistical unit.

2.7. Degree of Conversion

The measurements were made using FT-Raman spectroscopy (Spectrum GX spectrometer, PerkinElmer, Waltham, MA, USA). The excitation source was a NdYaG laser

(1064 nm), kept at 400 mW laser power and resolution of 8 cm^{-1}. For each spectrum, 300 scans were recorded.

DC was measured on the top and bottom surfaces of the specimens that were previously stored in saline solution for 1 day (n = 5) to determine the mechanical properties. Spectra of the uncured composites (n = 5) were acquired correspondingly. Kinetics add-on for MATLAB (Mathworks, Natick, MA, USA) was used for spectral analysis.

DC calculation was performed by comparing the peak heights of aliphatic C–C stretching mode at 1638 cm^{-1} and aromatic C$\cdots$C band at 1608 cm^{-1} attained from cured and uncured specimens using the following equation:

$$\text{DC (\%)} = \left(1 - \frac{(1638\ \text{cm}^{-1}/1608\ \text{cm}^{-1})\ \text{after curing}}{(1638\ \text{cm}^{-1}/1608\ \text{cm}^{-1})\ \text{before curing}}\right) \times 100\ \% \quad (1)$$

2.8. Statistical Analysis

Mean values of FS, FM, MH, and DC were compared among the experimental materials using one-way ANOVA with Tukey's post-hoc adjustment for multiple comparisons. Pairwise comparisons (between 1 and 28 days for FS and FM; between 0 and 2 mm for the DC) were compared using a two-tailed t-test for independent samples. The t-test for independent samples was used for pairwise comparisons between 1 and 28 days for MH; p-values lower than 0.05 were considered statistically significant. The statistical analysis was performed using SPSS (version 20, IBM, Armonk, NY, USA).

3. Results

3.1. X-ray Diffraction

Figure 1 shows the amorphous structure of both bioactive glasses used in the study. Cu-MBGN showed a typical XRD pattern of amorphous silicate materials, in which only a broad band located at 2θ = 23° was observed [24].

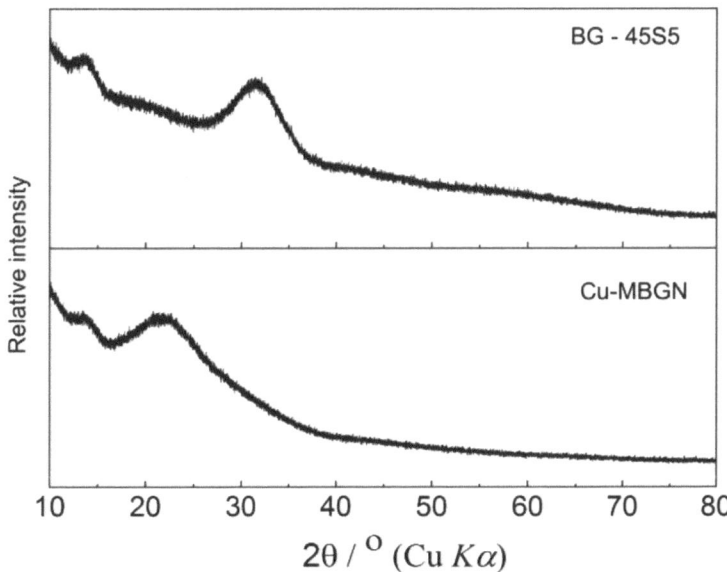

Figure 1. XRD spectra of conventional 45S5 bioactive glass (BG-45S5) and copper-doped mesoporous bioactive glass nanospheres (Cu-MBGN).

3.2. Scanning Electron Microscopy

Figure 2 shows a typical morphology of bioactive glass used in experimental materials. Conventional 45S5 BG had an irregular structure with a wide dispersion of particle sizes ranging from 200 nm up to 10 μm, with an average of 2–3 μm. Cu-MBGN presented uniform spherical particles that were approximately 100 nm in diameter.

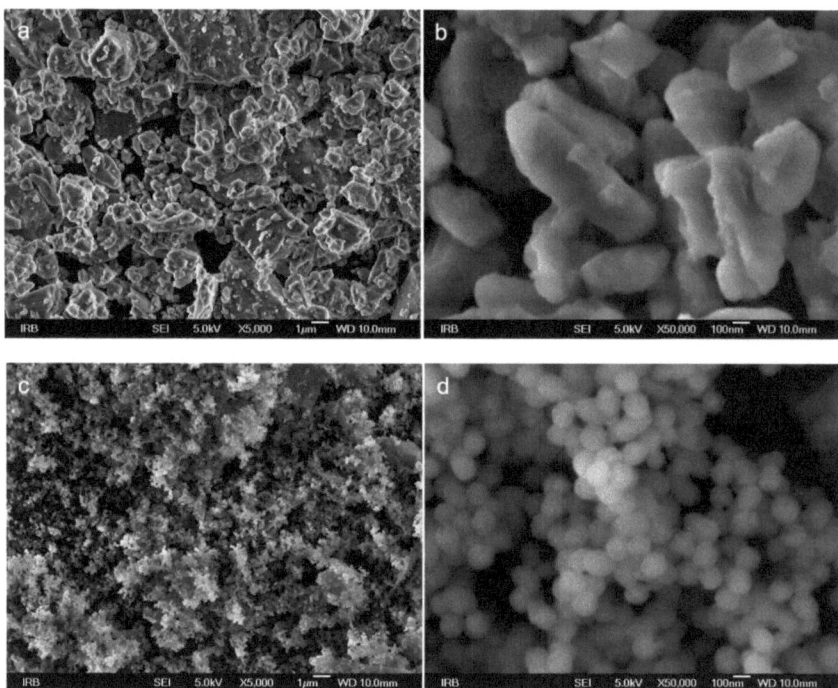

Figure 2. SEM of individual filler particles at 5000× and 50,000× magnification: (**a**,**b**) 45S5 bioactive glass; (**c**,**d**) Cu-MBGN.

3.3. Flexural Strength

The results shown in Figure 3 indicate that 1-CuBG-Si had the highest FS among Cu-MBGN-containing materials, both after 1 day and after 28 days. Control materials 10-BG and 10-Si had similarly high values at 1 day but saw a significant drop after 28 days of exposure to saline solution. The combination of Cu-MBGN and silica also seemed to be beneficial as there was no difference between 1 and 28 days for 1-CuBG-Si, 5-CuBG-Si, and 10-CuBG-Si. On the other hand, materials with Cu-MBGN only (10-CuBG), BG only (10-BG and 15-BG), or silica only (10-Si and 15-Si) did not demonstrate the same stability.

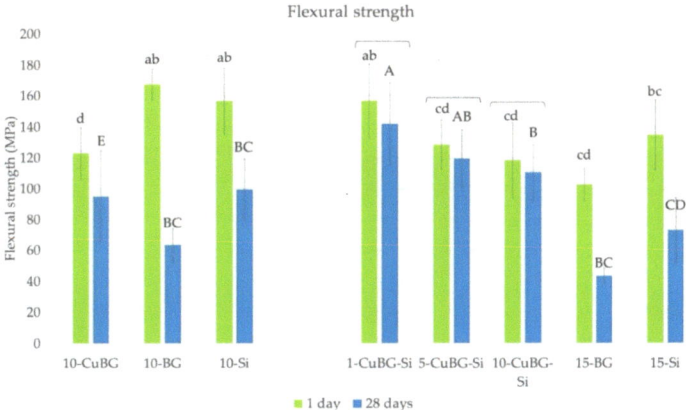

Figure 3. Flexural strength of the tested materials. Lower case letters indicate no statistically significant differences between groups after 1 day. Upper case letters indicate no statistically significant differences between groups after 28 days. Brackets indicate no statistically significant difference within the same material between 1 and 28 days.

3.4. Flexural Modulus

The results are depicted in Figure 4. Again, materials containing both Cu-MBGN and silica had the highest modulus after 1 day, regardless of their amount. However, there was no difference from the inert control 15-Si. After 28 days, 5-CuBG-Si was the material with the highest values. In contrast, materials with 10 and 15% of regular bioactive glass without silica (10-BG and 15-BG) had the lowest modulus.

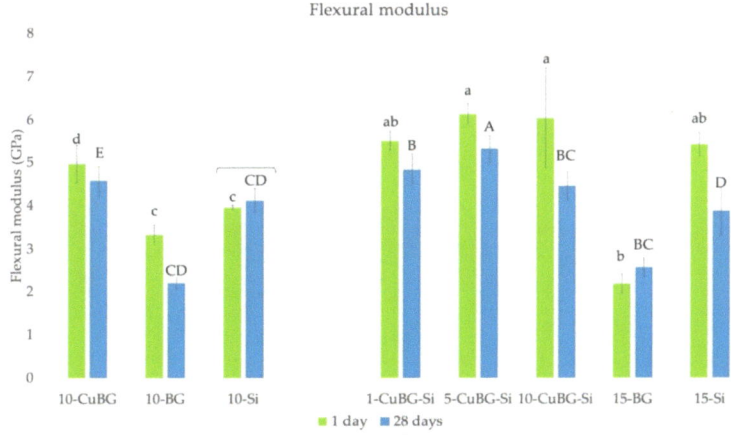

Figure 4. Flexural modulus of the tested materials. Lower case letters indicate no statistically significant differences between groups after 1 day. Upper case letters indicate no statistically significant differences between groups after 28 days. Brackets indicate no statistically significant difference within the same material between 1 and 28 days.

3.5. Microhardness

The MH (Figure 5) was generally equal or higher for CuBG composites than the corresponding controls after the 28 days of exposure to saline solution. Materials 1-CuBG-Si and 10-CuBG-Si had significantly higher MH after 28 days. Material 10-CuBG was the only one with a significant decrease in MH after prolonged contact with the aqueous environment.

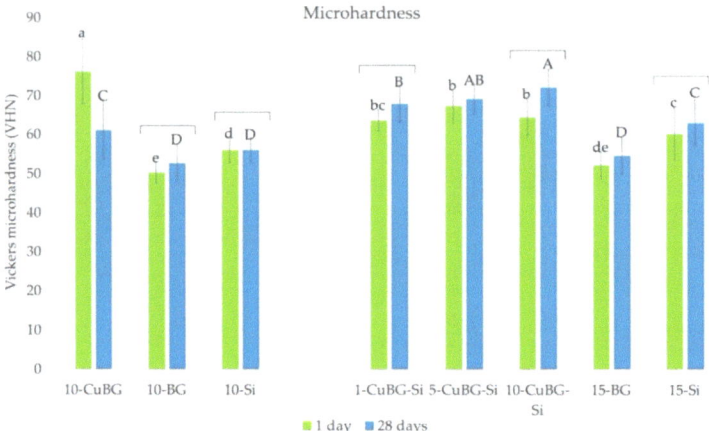

Figure 5. Vickers microhardness (VHN) of the tested materials after 1 and 28 days of exposure to saline solution measured at the top of the specimen. Lower case letters indicate no statistically significant differences between groups after 1 day. Upper case letters indicate no statistically significant differences between groups after 28 days. Brackets indicate no statistically significant difference within the same material between 1 and 28 days.

3.6. Degree of Conversion

The results are shown in Figure 6. All the materials, except 10-CuBG-Si, had a very high DC of above 80% at the top (0 mm) and bottom (2 mm) surfaces. There were no differences between the top and bottom surfaces for all materials. The exception was 10-CuBG-Si, which showed the lowest top DC, while the DC at 2 mm was statistically similar to other Cu-MBGN materials with silica.

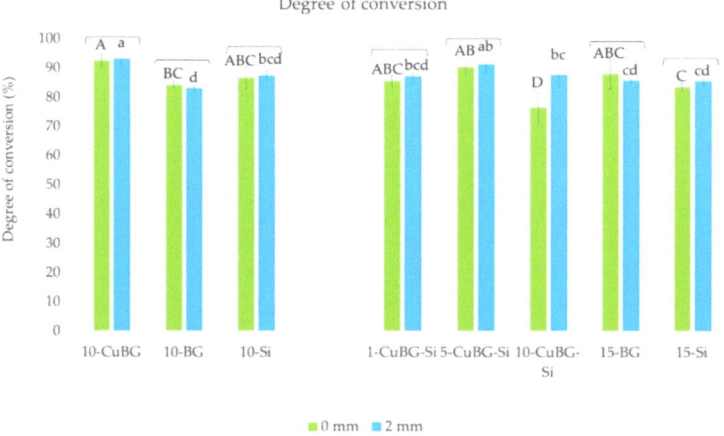

Figure 6. Degree of conversion of the tested materials 1 day after polymerization and exposure to saline solution. Upper case letters indicate no statistically significant differences between groups at the top (0 mm) surface. Lower case letters indicate no statistically significant differences between groups at the bottom (2 mm) surface. Brackets indicate no statistically significant difference within the same material between the top and bottom surfaces.

4. Discussion

Incorporation of antimicrobial and remineralizing components into a dental resin composite could be an effective strategy for preventing secondary caries. In this study, experimental dental composites with Cu-MBGN were developed to mitigate this issue. Trimodal approach with the addition of Cu-MBGN combined with silica nanofillers and inert glass microfillers seemed to benefit the durability of their FS and MH, especially after 28 days of exposure to saline solution. Conversely, FS, FM, and MH of the corresponding control materials containing 45S5 bioactive glass with irregular microsized particles were the lowest at the same time point. Both null hypotheses were rejected.

4.1. Bimodal Approach

The design of a bioactive dental restorative material is challenging. Demanding oral conditions of repeated masticatory load and constant changes in pH, temperature, chemical erosion, and enzymatic degradation require a mechanically resistant and stable material. On the other hand, bioactive materials interact with living tissue and release certain substances. Thus, material deterioration over time, at least to some extent, is expected. The balance between these opposites is delicate, and the materials should always be tested after intermediate or long-term exposure to an aqueous environment similar to saliva. To minimize unwanted material degradation, stable inert fillers with reinforcing properties were combined with bioactive fillers. Owing to the large reactive surface of Cu-MBGN, bioactivity is expected even when introduced in smaller quantities, allowing five times higher share of inert fillers. In this study, we opted to compare the sole effect of Cu-MBGN particles on the mechanical and curing properties of composite bimodal mixtures to inert and commercial bioactive control before examining the combined effect of Cu-MBGN and silica nanofillers in a trimodal approach.

Bimodal 10-CuBG without silica fillers showed lower FS than the corresponding inert (10-Si) and bioactive control (10-BG). The strength of the resin composites is mainly controlled by the amount of fillers, their size, and the quality of their bond to resin [30]. The mesoporous nanoparticles used here were characterized by a large surface area and small volume but no silanization. The lack of filler/matrix bond most likely interrupted effective stress transfer (between fillers and resin) upon specimen loading. While the addition of silanized fillers improves the FS, the addition of unsilanized fillers reduces it [30]. Similar results were obtained in a study that examined the effect of mesoporous silica nanoparticles on acrylic bone cement [31] and poly(methyl methacrylate) for denture base [32], where the increase in the unsilanized filler load decreased the FS. Conversely, silanized zinc-doped mesoporous silica nanoparticles increased the FS of the experimental dental resins in a simple unimodal filler mixture with a maximum 15 wt % total filler loading [33].

Our SEM analysis showed a significant difference in the shape and size of two bioactive glasses used in the present study. Irregular shape of 45S5 BG is typical for a melt-quench method, while microemulsion-assisted sol–gel method developed by Zheng et al. [24] typically produces spherical particles with diameter of 100–300 nm, as demonstrated in this study. Expectedly, both bioactive fillers showed amorphous structure in XRD analysis. Despite being unsilanized, commercial bioactive glass fillers in 10-BG were significantly larger (4 µm) and thus had a lower contact surface area to resin. Therefore, there were fewer crack initiation sites, which led to a short-term favorable FS in 10-BG. However, the well-known water degradation of 45S5 BG caused the FS of 10-CuBG to surpass the FS of 10-BG after 28 days of exposure to saline solution.

Unlike FS, the FM and MH were enhanced by introducing Cu-MBGN, manifested by the higher values for 10-CuBG than 10-Si and 10-BG. The increase in FM is directly correlated to the filler loading but independent of the interfacial adhesion [30]. Generally, the addition of fillers elevates the FM of the resin matrix because most fillers have greater stiffness than organic polymers [30]. Similarly, an increase in the filler volume and DC improves the MH [34]. The higher filler volume of 10-CuBG compared to that of 10-Si explains our results, similar to other studies [18,31,32].

However, higher filler volume cannot explain the low FM of 10-BG specimens. Commercial BG fillers previously showed a dose-dependent decrease in FM and FS in composites with identical resin matrix and fillers and similar filler amounts as in the present study [8]. High hydrophilicity of 45S5 particles and concomitant water sorption, along with the lack of interfacial bonding and low DC, were identified as culprits for poor mechanical properties. The same dose dependency for FS and FM was seen in this study.

In conclusion, the incorporation of Cu-MBGN in bimodal composite mixtures led to improvement of polymerization, stiffness, and hardness but degradation of strength.

4.2. Trimodal Approach

To control the strength reduction caused by the Cu-MBGN addition, further refinement and adaptation of the composite structure was necessary. The majority of contemporary commercial nanofilled composite materials consist of microfillers with a minor share of nanofillers. Smaller particles occupy the space between larger particles and increase the material's packing density [26]. This approach was also used here to achieve a higher filler amount, with two sizes of nanoparticles, namely 12 nm (silica) and ~100 nm (Cu-MBGN), along with inert glass microfillers.

The trimodal filler mixture was apparently more successful than the bimodal approach. Cu-BG composites with silica (1-CuBG-Si, 5-CuBG-Si, and 10-CuBG-Si) kept high FS values after 28 days and were statistically similar to values after 1 day. The reinforcing effect of Cu-MBGN particles could be ascribed to a combination of resin interlocking in the porosities of Cu-MBGN and additional reinforcing effect of silica, which prevented composite deterioration in water after 28 days. On the other hand, a drop of FS in the inert control 15-Si could be related to silane water degradation, leading to hydrolytic degradation and resin plasticization [13].

Similar to the bimodal approach, in CuBG composites with silica, the gradual replacement of silica with Cu-MBGN also provided a higher volume and higher FM and MH, thus retaining the favorable stiffness and hardness. The results are in agreement with other studies [18,31,32].

All the tested materials achieved high DC values except for 10-CuBG-Si. The high surface area of mesoporous nanoparticles limits their addition to resin. A higher amount of resin is necessary to wet each particle, leading to a steep increase in the composite's viscosity with higher Cu-MBGN inclusion. Uniform dispersion of nanoparticles becomes increasingly difficult with higher amount. If unsuccessful, particles group together due to weak van der Waals attractive forces [26,31]. Material 10-CuBG-Si had the highest volume filler loading in this study. Furthermore, the material had poor handling properties and was visually very dry. Significant packing was necessary to produce satisfactory specimens. The consequences of this disadvantage were not apparent in the tested properties, except for DC. The 35 wt % of resin was not sufficient to ensure adequate radical mobility and complete polymerization of 10-CuBG-Si. In contrast, the favorable viscosity of bimodal 10-CuBG probably contributed to the highest DC of all the tested materials. Unsilanized Cu-MBGN did not limit the mobility of the radicals, leading to a higher number of reacted C=C bonds [35].

A key to uniform dispersion of nanoparticles in general, and especially mesoporous particles, in resin composites is the mixing process. Numerous techniques have been developed to overcome this obstacle [26,31]. A high-temperature, high vacuum mixing process was proposed by Samuel et al. [18]. An unfilled resin matrix was heated to 80 °C to reduce its viscosity and achieve penetration of organic monomers into porosities at the surface of filler particles. Additionally, vacuum was used to release trapped air inclusions and facilitate the flow into porosities. In that study, mesoporous particles had relatively large porosities of 0.46 cm^3/g with a 50% volume fraction [18]. Although the resin was preheated in this study, there is still room for improvement.

Trimodal filler mixture, in particular materials with 1 and 5% of Cu-MBGN, demonstrated high polymerization, strength, modulus, and hardness.

5. Conclusions

The bimodal combination of Cu-MBGN and microfillers elevated the stiffness and hardness of the resin composites while decreasing strength. However, trimodal CuBG composites with silica and microfillers demonstrated durable strength after 28 days and increased elastic modulus and hardness, with very high polymerization efficiency. Due to these characteristics, the trimodal filler mixture appears to be a promising path for future investigations of antibacterial properties and ion release. Within the limitations of this study, we expect that the incorporation of Cu-MBGN as an addition to inert silica and glass microfillers in dental resin composite could result in an effective and resistant remineralizing restorative material.

Author Contributions: Conceptualization, D.M., H.J.H., M.P. and A.R.B.; data curation, D.M., V.N.M. and K.Z.; formal analysis, V.N.M. and M.P.; investigation, D.M., V.N.M. and K.Z.; methodology, D.M., H.J.H., M.P., V.N.M. and Z.T.; project administration, D.M.; resources, H.J.H., K.Z., Z.T. and A.R.B.; software, M.P.; supervision, H.J.H., Z.T. and A.R.B.; validation, Z.T., D.M. and M.P.; visualization, M.P.; writing—original draft, D.M.; writing—review and editing, H.J.H., V.N.M., M.P., K.Z., Z.T. and A.R.B. All authors have read and agreed to the published version of the manuscript.

Funding: This study was fully supported by the Croatian Science Foundation, grant number IP-2019-04-6183.

Institutional Review Board Statement: Not applicable.

Informed Consent Statement: Not applicable.

Data Availability Statement: The datasets generated and analyzed in the present study are available from the corresponding author on reasonable request.

Acknowledgments: The authors gratefully acknowledge the donation of fillers from Evonik and Schott and the curing unit from Ivoclar Vivadent.

Conflicts of Interest: The authors declare no conflict of interest. The funders had no role in the design of the study; in the collection, analyses, or interpretation of data; in the writing of the manuscript, or in the decision to publish the results.

References

1. Opdam, N.J.; van de Sande, F.H.; Bronkhorst, E.; Cenci, M.S.; Bottenberg, P.; Pallesen, U.; Gaengler, P.; Lindberg, A.; Huysmans, M.C.; van Dijken, J.W. Longevity of posterior composite restorations: A systematic review and meta-analysis. *J. Dent. Res.* **2014**, *93*, 943–949. [CrossRef] [PubMed]
2. Kidd, E.A. Diagnosis of secondary caries. *J. Dent. Educ.* **2001**, *65*, 997–1000. [CrossRef] [PubMed]
3. Marovic, D.; Tarle, Z.; Hiller, K.A.; Muller, R.; Rosentritt, M.; Skrtic, D.; Schmalz, G. Reinforcement of experimental composite materials based on amorphous calcium phosphate with inert fillers. *Dent. Mater.* **2014**, *30*, 1052–1060. [CrossRef] [PubMed]
4. Marovic, D.; Tarle, Z.; Hiller, K.A.; Muller, R.; Ristic, M.; Rosentritt, M.; Skrtic, D.; Schmalz, G. Effect of silanized nanosilica addition on remineralizing and mechanical properties of experimental composite materials with amorphous calcium phosphate. *Clin. Oral Investig.* **2014**, *18*, 783–792. [CrossRef]
5. Marovic, D.; Sariri, K.; Demoli, N.; Ristic, M.; Hiller, K.A.; Skrtic, D.; Rosentritt, M.; Schmalz, G.; Tarle, Z. Remineralizing amorphous calcium phosphate based composite resins: The influence of inert fillers on monomer conversion, polymerization shrinkage, and microhardness. *Croat. Med. J.* **2016**, *57*, 465–473. [CrossRef]
6. Par, M.; Santic, A.; Gamulin, O.; Marovic, D.; Mogus-Milankovic, A.; Tarle, Z. Impedance changes during setting of amorphous calcium phosphate composites. *Dent. Mater.* **2016**, *32*, 1312–1321. [CrossRef]
7. Par, M.; Spanovic, N.; Bjelovucic, R.; Marovic, D.; Schmalz, G.; Gamulin, O.; Tarle, Z. Long-term water sorption and solubility of experimental bioactive composites based on amorphous calcium phosphate and bioactive glass. *Dent. Mater. J.* **2019**, *38*, 555–564. [CrossRef]
8. Par, M.; Tarle, Z.; Hickel, R.; Ilie, N. Mechanical properties of experimental composites containing bioactive glass after artificial aging in water and ethanol. *Clin. Oral Investig.* **2019**, *23*, 2733–2741. [CrossRef]
9. Par, M.; Tarle, Z.; Hickel, R.; Ilie, N. Dentin bond strength of experimental composites containing bioactive glass: Changes during aging for up to 1 Year. *J. Adhes. Dent.* **2018**, *20*, 325–334. [CrossRef]

10. Par, M.; Tarle, Z.; Hickel, R.; Ilie, N. Polymerization kinetics of experimental bioactive composites containing bioactive glass. *J. Dent.* **2018**, *76*, 83–88. [CrossRef]
11. Par, M.; Spanovic, N.; Bjelovucic, R.; Skenderovic, H.; Gamulin, O.; Tarle, Z. Curing potential of experimental resin composites with systematically varying amount of bioactive glass: Degree of conversion, light transmittance and depth of cure. *J. Dent.* **2018**, *75*, 113–120. [CrossRef]
12. Langhorst, S.E.; O'Donnell, J.N.; Skrtic, D. In vitro remineralization of enamel by polymeric amorphous calcium phosphate composite: Quantitative microradiographic study. *Dent. Mater.* **2009**, *25*, 884–891. [CrossRef]
13. Xavier, T.A.; Fróes-Salgado, N.R.D.G.; Meier, M.M.; Braga, R.R. Influence of silane content and filler distribution on chemical-mechanical properties of resin composites. *Braz. Oral Res.* **2015**, *29*, 1–8. [CrossRef]
14. Skrtic, D.; Antonucci, J.M.; Eanes, E.D.; Eidelman, N. Dental composites based on hybrid and surface-modified amorphous calcium phosphates. *Biomaterials* **2004**, *25*, 1141–1150. [CrossRef]
15. Tiskaya, M.; Al-Eesa, N.A.; Wong, F.S.L.; Hill, R.G. Characterization of the bioactivity of two commercial composites. *Dent. Mater.* **2019**, *35*, 1757–1768. [CrossRef]
16. Porenczuk, A.; Jankiewicz, B.; Naurecka, M.; Bartosewicz, B.; Sierakowski, B.; Gozdowski, D.; Kostecki, J.; Nasiłowska, B.; Mielczarek, A. A comparison of the remineralizing potential of dental restorative materials by analyzing their fluoride release profiles. *Adv. Clin. Exp. Med.* **2019**, *28*, 815–823. [CrossRef]
17. Atai, M.; Pahlavan, A.; Moin, N. Nano-porous thermally sintered nano silica as novel fillers for dental composites. *Dent. Mater.* **2012**, *28*, 133–145. [CrossRef]
18. Samuel, S.P.; Li, S.; Mukherjee, I.; Guo, Y.; Patel, A.C.; Baran, G.; Wei, Y. Mechanical properties of experimental dental composites containing a combination of mesoporous and nonporous spherical silica as fillers. *Dent. Mater.* **2009**, *25*, 296–301. [CrossRef]
19. Bari, A.; Bloise, N.; Fiorilli, S.; Novajra, G.; Vallet-Regí, M.; Bruni, G.; Torres-Pardo, A.; González-Calbet, J.M.; Visai, L.; Vitale-Brovarone, C. Copper-containing mesoporous bioactive glass nanoparticles as multifunctional agent for bone regeneration. *Acta Biomater.* **2017**, *55*, 493–504. [CrossRef]
20. Hench, L.L. The story of Bioglass. *J. Mater. Sci. Mater. Med.* **2006**, *17*, 967–978. [CrossRef]
21. Zheng, K.; Wu, J.; Li, W.; Dippold, D.; Wan, Y.; Boccaccini, A.R. Incorporation of cu-containing bioactive glass nanoparticles in gelatin-coated scaffolds enhances bioactivity and osteogenic activity. *ACS Biomater. Sci. Eng.* **2018**, *4*, 1546–1557. [CrossRef]
22. Zheng, K.; Sui, B.; Ilyas, K.; Boccaccini, A.R. Porous bioactive glass micro- and nanospheres with controlled morphology: Developments, properties and emerging biomedical applications. *Mater. Horiz.* **2021**, *8*, 300–335. [CrossRef]
23. Yang, Y.; Zheng, K.; Liang, R.; Mainka, A.; Taccardi, N.; Roether, J.A.; Detsch, R.; Goldmann, W.H.; Virtanen, S.; Boccaccini, A.R. Cu-releasing bioactive glass/polycaprolactone coating on Mg with antibacterial and anticorrosive properties for bone tissue engineering. *Biomed. Mater.* **2017**, *13*, 015001. [CrossRef]
24. Zheng, K.; Kang, J.; Rutkowski, B.; Gawęda, M.; Zhang, J.; Wang, Y.; Founier, N.; Sitarz, M.; Taccardi, N.; Boccaccini, A.R. Toward highly dispersed mesoporous bioactive glass nanoparticles with high cu concentration using cu/ascorbic acid complex as precursor. *Front. Chem.* **2019**, *7*, 497. [CrossRef]
25. Rau, J.V.; Curcio, M.; Raucci, M.G.; Barbaro, K.; Fasolino, I.; Teghil, R.; Ambrosio, L.; De Bonis, A.; Boccaccini, A.R. Cu-releasing bioactive glass coatings and their in vitro properties. *ACS Appl. Mater. Interfaces* **2019**, *11*, 5812–5820. [CrossRef] [PubMed]
26. Jandt, K.D.; Watts, D.C. Nanotechnology in dentistry: Present and future perspectives on dental nanomaterials. *Dent. Mater.* **2020**, *36*, 1365–1378. [CrossRef]
27. Rivera, L.R.; Cochis, A.; Biser, S.; Canciani, E.; Ferraris, S.; Rimondini, L.; Boccaccini, A.R. Antibacterial, pro-angiogenic and pro-osteointegrative zein-bioactive glass/copper based coatings for implantable stainless steel aimed at bone healing. *Bioact. Mater.* **2021**, *6*, 1479–1490. [CrossRef]
28. Wu, C.; Zhou, Y.; Xu, M.; Han, P.; Chen, L.; Chang, J.; Xiao, Y. Copper-containing mesoporous bioactive glass scaffolds with multifunctional properties of angiogenesis capacity, osteostimulation and antibacterial activity. *Biomaterials* **2013**, *34*, 422–433. [CrossRef]
29. Ilie, N.; Hickel, R. Investigations on mechanical behaviour of dental composites. *Clin. Oral Investig.* **2009**, *13*, 427–438. [CrossRef]
30. Fu, S.-Y.; Feng, X.-Q.; Lauke, B.; Mai, Y.-W. Effects of particle size, particle/matrix interface adhesion and particle loading on mechanical properties of particulate–polymer composites. *Compos. Part B Eng.* **2008**, *39*, 933–961. [CrossRef]
31. Slane, J.; Vivanco, J.; Meyer, J.; Ploeg, H.L.; Squire, M. Modification of acrylic bone cement with mesoporous silica nanoparticles: Effects on mechanical, fatigue and absorption properties. *J. Mech. Behav. Biomed. Mater.* **2014**, *29*, 451–461. [CrossRef] [PubMed]
32. Lee, J.H.; El-Fiqi, A.; Jo, J.K.; Kim, D.A.; Kim, S.C.; Jun, S.K.; Kim, H.W.; Lee, H.H. Development of long-term antimicrobial poly(methyl methacrylate) by incorporating mesoporous silica nanocarriers. *Dent. Mater.* **2016**, *32*, 1564–1574. [CrossRef] [PubMed]
33. Bai, X.; Lin, C.; Wang, Y.; Ma, J.; Wang, X.; Yao, X.; Tang, B. Preparation of Zn doped mesoporous silica nanoparticles (Zn-MSNs) for the improvement of mechanical and antibacterial properties of dental resin composites. *Dent. Mater.* **2020**, *36*, 794–807. [CrossRef] [PubMed]
34. Neves, A.D.; Discacciati, J.A.; Orêfice, R.L.; Jansen, W.C. Correlation between degree of conversion, microhardness and inorganic content in composites. *Pesqui. Odontol. Bras.* **2002**, *16*, 349–354. [CrossRef]
35. Wilson, K.S.; Zhang, K.; Antonucci, J.M. Systematic variation of interfacial phase reactivity in dental nanocomposites. *Biomaterials* **2005**, *26*, 5095–5103. [CrossRef]

Article

Nanoparticles of Bioactive Glass Enhance Biodentine Bioactivity on Dental Pulp Stem Cells

Camila Corral Nunez [1,*], Diego Altamirano Gaete [1], Miguel Maureira [2], Javier Martin [1] and Cristian Covarrubias [2,*]

1. Department of Restorative Dentistry, Faculty of Dentistry, Universidad de Chile, Santiago 8380544, Chile; dgaete1892@gmail.com (D.A.G.); jamartin@odontologia.uchile.cl (J.M.)
2. Laboratory of Nanobiomaterials, Research Institute of Dental Sciences, Faculty of Dentistry, University of Chile, Santiago 8380544, Chile; miguel.maureira@bq.uchile.cl
* Correspondence: camila.corral@odontologia.uchile.cl (C.C.N.); ccovarrubias@odontologia.uchile.cl (C.C.); Tel.: +56-2-9781742 (C.C.N.); +56-2-9785063 (C.C.)

Abstract: This study aimed to investigate the cytotoxicity and bioactivity of a novel nanocomposite containing nanoparticles of bioactive glass (nBGs) on human dental pulp stem cells (hDPSCs). nBGs were synthesized by the sol–gel method. Biodentine (BD) nanocomposites (nBG/BD) were prepared with 2 and 5% wt of nBG content; unmodified BD and glass ionomer cement were used as references. Cell viability and attachment were evaluated after 3, 7 and 14 days. Odontogenic differentiation was assessed with alkaline phosphatase (ALP) activity after 7 and 14 days of exposure. Cells successfully adhered and proliferated on nBG/BD nanocomposites, cell viability of nanocomposites was comparable with unmodified BD and higher than GIC. nBG/BD nanocomposites were, particularly, more active to promote odontogenic differentiation, expressed as higher ALP activity of hDPSCs after 7 days of exposure, than neat BD or GIC. This novel nanocomposite biomaterial, nBG/BD, allowed hDPSC attachment and proliferation and increased the expression of ALP, upregulated in mineral-producing cells. These findings open opportunities to use nBG/BD in vital pulp therapies.

Keywords: apatite-forming ability; bioactive glass; bioactivity; nanocomposites

Citation: Corral Nunez, C.; Altamirano Gaete, D.; Maureira, M.; Martin, J.; Covarrubias, C. Nanoparticles of Bioactive Glass Enhance Biodentine Bioactivity on Dental Pulp Stem Cells. *Materials* **2021**, *14*, 2684. https://doi.org/10.3390/ma14102684

Academic Editors: Tobias Tauböck and Matej Par

Received: 15 March 2021
Accepted: 20 April 2021
Published: 20 May 2021

Publisher's Note: MDPI stays neutral with regard to jurisdictional claims in published maps and institutional affiliations.

Copyright: © 2021 by the authors. Licensee MDPI, Basel, Switzerland. This article is an open access article distributed under the terms and conditions of the Creative Commons Attribution (CC BY) license (https://creativecommons.org/licenses/by/4.0/).

1. Introduction

Current scientific evidence has provided support to treat pulpal exposures caused by dental trauma injuries or caries lesions with vital pulp therapies (VPTs) [1–4]. According to the recently published guidelines for dental trauma management, every effort should be made to preserve the vitality of this tissue in immature and mature teeth, recommending conservative pulpal therapy approaches [5]. Meanwhile, the European Society of Endodontology has opened the door for this paradigm shift, also recommending VPTs such as pulp capping and pulpotomy (partial and full) for cariously exposed pulps [6].

VPT aims to remove the microbial irritation and protect the exposed tissues from external stimuli by placing a well-sealing dental material. Ideally, this material should not be toxic to the pulp cells, but it should also be bioactive towards the tissues by stimulating migration, proliferation and odontogenic differentiation of the cells. Traditionally, materials based on calcium hydroxide or calcium silicate have been used for this purpose [7–9]. Calcium hydroxide, due to its alkaline pH, exhibits an antibacterial effect and induces superficial necrosis in the pulpal tissues. This is thought to promote odontoblast differentiation and the formation of a dentin bridge [9,10]. However, its main drawbacks are its dissolution over time, its lack of bonding to the dentin, leading to susceptibility to leakage, and the tunnel defects in the dentin bridge formed [7,11,12]. In the 1990s, mineral trioxide aggregate (MTA) appeared, a calcium silicate-based cement (CSC) commercially available as ProRoot MTA (Tulsa Dental Products, Tulsa, OK, USA), composed mainly of Portland cement and bismuth oxide [13,14]. MTA has proven excellent biocompatibility, with the ability to induce mineralization, showing high clinical success rates for VPT, generally inducing bridge formation [7,11,15,16]. However, its long setting time [17], tooth discoloration [18], high cost and difficult handling characteristics are considered its main disadvantages [8,11]. Subsequently, several other CSCs have been developed, which, depending on the purpose, can be classified into restorative cements used in VPT (Biodentine™, MTA Angelus, RetroMTA and TheraCal LC) and endodontic sealers (BioRoot RCS) [8]. Biodentine™ (Septodont, St. Maur-des-Fossés, France) has been manufactured to overcome the disadvantages of MTA [19], exhibiting reduced setting time [20], enhanced handling and mechanical properties [21] and adequate radiopacity [22]. It has also shown good clinical success in VPT, inducing complete dentinal bridge formation in exposed pulps of asymptomatic and symptomatic vital permanent teeth [8,23–26]. Moreover, evidence from in vitro and animal studies showed the capacity of Biodentine to promote greater mineralized tissue deposition than other CSCs [27,28]

Since the sealing capability of materials used in VPT is key, materials that bond to dentin, stimulate odontogenic differentiation of pulpal cells and induce rapid formation of a dentin bridge are ideal to allow a reliable sealing. For this purpose, bioceramic-based materials have mainly been explored, including traditional bioceramics such as hydroxyapatite [29], calcium phosphate [30], gelatine/hydroxyapatite/tricalcium phosphate composites [31] or novel sol–gel SiO_2/ZrO_2 ceramic composites [32]. In addition, bioactive glass (BG) is a bioceramic that exhibits advanced bioactive properties for VPT applications. The original BG, developed by Hench, is composed of 46.1 mol.% SiO_2, 24.4 mol.% Na_2O, 26.9 mol.% CaO and 2.6 mol.% P_2O_5, and it is known as 45S5 and Bioglass [33]. BG was initially used for medical applications due to its ability to bond to bone through the formation of an apatite layer on its surface [34,35]. However, it has also since been used in dentistry, incorporated into toothpaste for enamel remineralization and used for the treatment of dentin hypersensitivity [33]. In addition, BG incorporation into other dental materials, such as endodontic sealers [36,37], resin adhesives [38,39] and resin composites [40–43], has been explored. This was the topic of a recent critical review, which concludes that the addition of BG into dental composites is promising, presenting multiple benefits, especially its capacity to promote the precipitation of apatite [44]. When incorporated to an endodontic sealer, it has demonstrated in vitro capacity to promote differentiation of human periodontal ligament stem cells into cementoblast-like cells, enhancing the expression of genes related to the production of mineralized tissues [36]. Additionally, when this material was tested in a subcutaneous implantation model, it evidenced an adequate tissue reaction [37]. BG incorporated to resin adhesives showed the ability to bond and remineralize dentin [38]. It also promoted the precipitation of hydroxyapatite and calcium carbonate, improving the hybrid layer stability [39]. Similarly, experimental resin composites with BG have demonstrated multiple advantages, including the remineralization of adjacent demineralized dentin [40], acid-neutralizing properties [45], a local antimicrobial effect [41], reduction of biofilm penetration into marginal gaps [42] and enhancement of their mechanical properties [46,47].

The use of nanoparticles of BG (nBGs) with a high surface-to-volume ratio is of tremendous interest because of their larger specific surface area and enhanced bioactivity compared to the micrometric-sized particles of BG [48,49]. In addition, nBGs exhibit a higher remineralization rate of dentin [50] and an antimicrobial effect [51] when compared to microsized BG. Furthermore, resin composites containing nBGs promote the formation of a more uniform apatite layer and improve their capability to increase pH when compared to resin composites containing microsized BG [52]. nBGs form apatite in contact with the physiological fluids and they have been proven capable of inducing differentiation into a mineralizing lineage of stem cells [49,53,54]. In rat dental pulp stem cells, they increase the expression of odontogenic-related genes and the capacity for mineralization [54], and in hDPSC, they increase ALP activity, osteocalcin (OC) and dentin sialophosphoprotein (DSPP) production, and the formation of mineralized nodules [53].

The specific use of BG in VPT has been explored in animal models, showing the formation of reparative dentin and only mild inflammatory response in pulp capping procedures [55–57]. More recently, in a clinical trial on primary teeth, it demonstrated the formation of a dentin bridge [58]. However, there are no commercially available materials with BG currently, nor with nBGs for VPT application. For more practical reasons, this material in its powder form is not convenient to be applied as a VPT material. The materials for VPT should have good handling properties to allow the correct placement in the constricted space where pulpal exposure may occur [8]. Therefore, in general, materials used for this application are applied when recently mixed, before setting, to later achieve higher mechanical properties [8,59].

The incorporation of nBGs into BD (nBG/BD) is a possible new nanocomposite material that could integrate the handling properties and the ability to set of BD, together with its mechanical, biocompatible and bioactive characteristics, with further increased bioactivity provided by the incorporation of the nBGs. It has been previously shown that this nanocomposite presents enhanced bioactive properties in simulated body fluid, allowing a faster deposition of apatite on the surface of the material [60]. However, its ability to sustain human dental pulp stem cell (hDPSC) viability and differentiation remains largely unknown. The cellular response to the material is particularly relevant, especially its ability to induce cellular differentiation into a mineralizing lineage, since the formation of dentin in the injured area is clinically desirable. In this context, the development of this bioactive material, for dentin–pulp complex regeneration, is an interesting perspective. Therefore, the aims of this work are:

- to assess the cytocompatibility, in terms of viability, adhesion and morphology of hDPSCs, on direct contact with nBG/BD.
- to assess the ability of nBG/BD to stimulate the differentiation of hDPSCs into a mineralizing lineage.

The null hypothesis was that there would be no difference in cytocompatibility and ability to stimulate differentiation of hDPSCs into a mineralizing lineage between nBG/BD nanocomposites and BD.

2. Materials and Methods

2.1. Bioactive Glass Nanoparticle Synthesis and Nanocomposite Preparation

nBG particles (size ca. 40–70 nm) were synthesized by the sol–gel method, using the following molar composition: $58SiO_2:40CaO:5P_2O_5$ [61]. Briefly, a calcium-based solution was prepared by dissolving appropriate amounts of $Ca(NO_3)_2 \cdot 4H_2O$ (Merck, Darmstadt, Germany) in 120 mL of distilled water at room temperature. A second solution was prepared by diluting tetraethylorthosilicate (TEOS 98%; Sigma, Saint Louis, MO, USA) in 60 mL of ethanol, which was added to the calcium nitrate solution, and the pH of the resulting solution was adjusted to 2.0 with nitric acid. This transparent solution was slowly dropped under vigorous stirring into a solution of $NH_4H_2PO_4$ (May & Baker, Dagenham, England) in 1200 mL of distilled water. During the dripping process, the pH was kept at around 10 with aqueous ammonia. The reaction mixture was subjected to constant stirring for 48 h at 60 °C and allowed to stand for 24 h at room temperature. In this way, a precipitate was obtained, which was separated by centrifugation for 20 min at 12,000 rpm. This precipitate was washed through 3 cycles of centrifugation and redispersion of 40 min each. The solid obtained was frozen at −80 °C for 12 h, then lyophilized for 48 h and finally calcined at 700 °C for 3 h, producing a fine white powder of nBGs.

Nanocomposites based on nBG nanoparticles combined with BD (Septodont, Saint Maur des Fosses, France) were prepared. The 2% nBG/BD and 5% nBG/BD nanocomposites were obtained by adding 15 and 39 mg of nBG powder to the amount of BD existing in the commercial capsule, respectively. The resulting nBG/BD powders were then dry mixed within the BD capsule by using an amalgamator (Ultramat 2, SDI, Bayswater, Victoria, Australia) for 30 s. Five drops of BD liquid phase were then added to the capsule before mixing, according to the BD manufacturer's instructions. Neat BD and GIC (Fuji II, GC America Inc, Alsip, IL, USA) were used as reference materials and they were prepared following the manufacturer's instructions. Discs of the materials, measuring 6 mm in diameter and 1.5 mm thick, were prepared and allowed to fully set during incubation at 37 °C and 100% humidity for 24 h.

2.2. Nanocomposite Characterization

Discs of the nanocomposite materials (2% and 5% nBG/BD), BD and GIC were dehydrated, mounted on aluminum stubs and gold coated. Specimens were examined using scanning electron microscopy (SEM) in a JSM-IT300LV microscope (JEOL USA Inc., Peabody, MA, USA) equipped with an energy dispersive X-ray detector (EDX) and Aztec EDS software (Oxford Instruments, Abingdon, UK) for elemental analysis. SEM representative images at 1000× were obtained and EDX analysis of the surfaces of the materials was performed.

2.3. hDPSC Culture

The use of human cells in this study was approved by the Ethics Committee of the Faculty of Dentistry, University of Chile (Approval number PRI-ODO2018/06). Human dental pulp stem cells (hDPSCs) were isolated from human third molars, which were extracted for orthodontic reasons at the Dental Clinic, University of Chile. The extraction protocol was described by Covarrubias et al. [62].

2.4. hDPSC Viability Assays

hDPSCs were seeded directly onto the surface of 2% and 5% nBG/BD, BD and GIC discs (5 × 10^4 cells), placed in a single well of a 24-well cell culture plate and cultured in Dulbecco's modified Eagle medium (alpha-MEM; Invitrogen, Carlsbad, CA, USA) containing 10% fetal bovine serum (FBS, Gibco, Grand Island, NY, USA), 10 mM HEPES (Gibco, Grand Island, NY, USA), 100 U/mL penicillin and 100 mg/mL streptomycin (Sigma, Saint Louis, MO, USA). Cell viability was determined at 3, 7 and 14 days of incubation by using the CellTiter 96® Aqueous One Solution Cell Proliferation Assay (Promega, Madison, WI, USA), which measures the reduction of [3-(4,5-dimethylthiazol-2-yl)-5-(3-carboxymethoxyphenyl)-2-(4-sulfophenyl)-2H–tetrazolium] (MTS) to formazan by mitochondria in viable cells. Samples were incubated at 37 °C in a humidified 5% CO_2 atmosphere. The amount of soluble formazan produced by cellular reduction of MTS was measured by a microplate reader (InfiniteM200, Tecan, Crailsheim, Germany) at a wavelength of 490 nm.

2.5. hDPSC Morphology and Attachment

hDPSCs were directly seeded onto the material surfaces. After 7 and 14 days of incubation, the discs with cells were fixed (2% glutaraldehyde, Sigma-Aldrich, Saint Louis, MO, USA) and stored at 4 °C before starting the dehydration process. Discs were then immersed in increasing ethanol solutions. Critical point drying of specimens using CO_2 in an Autosamdri-815, Series A (Tousimis, Rockville, MD, USA) was performed. Samples were sputter-coated with 200 Å of gold and observed under SEM (Jeol JSM-IT300LV, JEOL USA Inc, Peabody, MA, USA). Representative micrographs of the surface of the materials after 7 and 14 days of culture were captured at 100× and 1000×.

2.6. Alkaline Phosphate Activity of hDPSCs

The capacity of the dental materials to stimulate the differentiation of hDPSCs into a mineralizing lineage was assessed by measuring the activity of the alkaline phosphatase (ALP) enzyme. ALP activity of hDPSCs cultured with the discs and a control (CT, without discs) was determined after 7 and 14 days of culture by a colorimetric dephosphorylation assay of a p-nitrophenyl phosphate reagent (Sigma, Saint Louis, MO, USA), which was followed by the measuring of the absorbance with a microplate reader (InfiniteM200, Tecan, Crailsheim, Germany) at a wavelength of 405 nm.

2.7. Statistical Analysis

Data obtained from the viability and alkaline phosphate activity assays were evaluated using SPSS software (SPSS Inc., Chicago, IL, USA). The results obtained for all materials were submitted to the Shapiro–Wilk normality test. After proving the normality of the sample data distribution, the data were submitted to a one- and two-way ANOVA and post hoc Tukey test at a 5% level of significance.

3. Results

3.1. Nanocomposite Characterization

SEM images and EDX elemental analysis of the surface of the set cements are presented in Figure 1. The surfaces of the nanocomposite discs (2% and 5% nBG/BD) appear similar to unmodified BD. BD, 2% nBG/BD and 5% nBG/BD are mainly composed (>10% by weight) of oxygen, calcium and carbon and a smaller amount (1–10% by weight) of silica and nitrogen, with traces (<1% by weight) of sodium and phosphorous. In contrast, the GIC surface presents visible cracks and it is composed mainly of oxygen, carbon and strontium and a smaller amount of aluminum and silica, with traces of sodium.

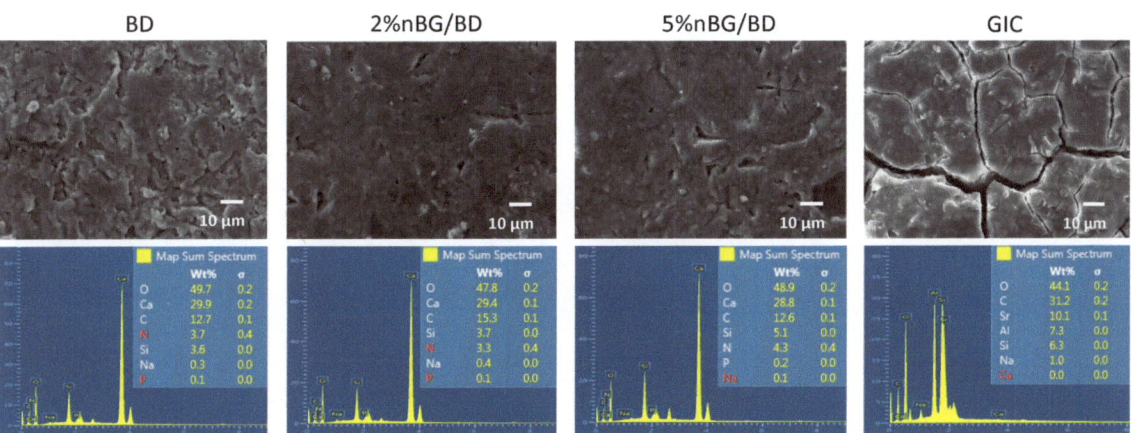

Figure 1. Representative SEM images (accelerating voltage of 20 kV, working distance of 10.3–10.6 mm, magnification of 1000×) and EDX elemental analysis of BD, 2% nBG/BD, 5% nBG/BD and GIC set cement surfaces.

3.2. hDPSC Viability

MTS cell viability of hDPSCs cultured in direct contact with the materials was assessed after 3, 7 and 14 days of incubation (Figure 2). The viability of hDPSCs cultured on 2% and 5% nBG/BD did not present statistical differences in comparison with those grown on unmodified BD. However, the viability of hDPCSs cultured with GIC was significantly lower than when cultured on BD and nanocomposites.

3.3. hDPSC Morphology and Adhesion

SEM images of hDPSCs attached to the cement surfaces are shown in Figure 3. After 7 days, cells covered almost the entire surface of the BD and nBG/BD materials, exhibiting a flattened morphology with cytoplasmic extensions projecting from the cells to adjacent cells. After 14 days of incubation, an increased density of cells and a thicker cell layer were observed. White and globular areas with the appearance of mineralized nodules could be also noted. In GIC samples, after 7 days of incubation, scarce cells with low adhesion were observed, and after 14 days, cells appeared more flattened and adhered although with notably lower cell density when compared to the other materials.

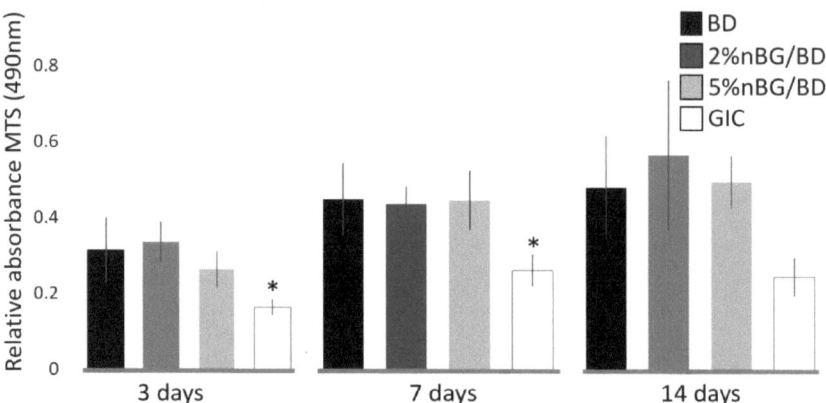

Figure 2. Cell viability of hDPSCSs cultured with BD, 2% nBG/BD, 5% nBG/BD and GIC at different culture times as determined by the MTS assay. Values are combined from 2 experiments ($n = 4$/experiment), standard deviations are represented by vertical bars. *: Statistically significant difference compared with BD ($p < 0.05$).

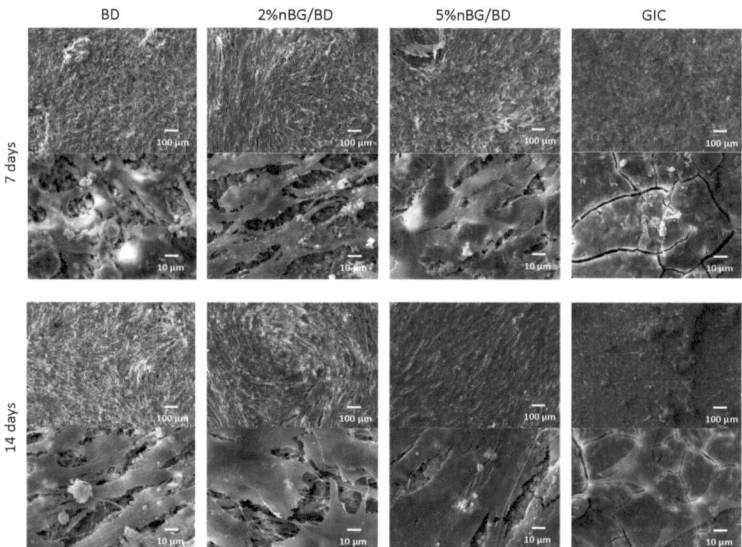

Figure 3. Representative SEM images (accelerating voltage of 20 kV, working distance of 10.0–11.4 mm, magnification 100× and 1000×) of hDPSCs cultured with BD, 2% nBG/BD, 5% nBG/BD and GIC after 7 and 14 days of incubation.

3.4. ALP Activity of hDPSCs

The ability of the nanocomposites to stimulate differentiation of hDPSCs towards a mineralizing lineage was assessed by quantifying the expression of ALP enzyme (Figure 4). Statistically, significantly higher ALP activity values were measured in hDPSCs in contact with 2 and 5% nBG/BD after 7 days of culture, when compared to hDPSCs in contact with the control (cells without material), BD and GIC. After 14 days of culture, the nanocomposites and BD presented a significantly higher ALP activity when compared to CT and GIC.

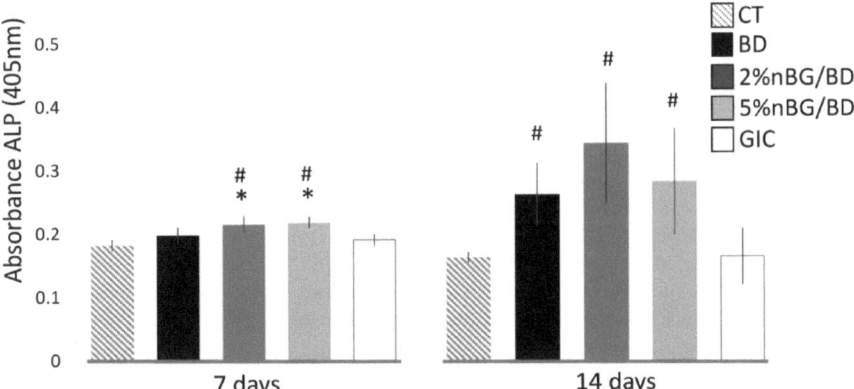

Figure 4. ALP activity of hDPSCs cultured with BD, 2% nBG/BD, 5% nBG/BD and GIC at different times. Values are means combined from 2 experiments ($n = 4$/experiment), standard deviations are represented by vertical bars. #: Statistically significant difference compared with control ($p < 0.05$). *: Statistically significant difference compared with BD ($p < 0.05$).

4. Discussion

In the present study, the incorporation of nBGs to a calcium silicate cement improved its ability to induce odontogenic differentiation, without generating significant changes in the adhesion and cell viability of hDPSCs.

The cytocompatibility of the nanocomposite materials was evaluated with the materials in direct contact with the hDPSC culture since this is the relationship that is established in VPTs. Cell viability measured through MTS mitochondrial activity and cell adhesion observations showed that the incorporation of nBGs in contents of 2 and 5% to BD does not affect the viability nor the adhesion of hDPSCs. To our best knowledge, no studies about cytocompatibility of nBG/BD composites have been reported. However, the viability of hDPSCs in contact with nBG-modified composites has been studied by means of different vehicles such as polymer hydrogels [63], synthetic polymers [64] and chitosan scaffolds [65]. In these studies, similarly to that observed in the current work, the presence of BG did not affect the adhesion and proliferation of the cells, with a cytocompatibility similar to the controls without BG.

SEM analysis of cells adhered to the materials revealed that hDPSCs behaved similarly when in contact with nBG/BD nanocomposites and BD. The cells were flattened, forming a well-organized layer covering the entire surface, with multiple extensions between the cells and towards the surface of the materials, indicating an effective cellular adhesion. In addition, the presence of nodules with a mineralizing appearance was also observed [62], which could be coupled to the differentiation process of the cells towards a mineralizing lineage. In contrast, the surface of the GIC was smooth and cracked, with poorly adhered cells, isolated and with a contracted appearance [66].

Cell differentiation into a mineralizing lineage was confirmed by determining the activity of ALP, an enzyme involved in the mineralizing process. hDPSCs cultured on the nBG/BD nanocomposites showed an early induction of ALP production after 7 days of culture, statistically higher than those incubated with neat BD. After 14 days, cells on nBG/BD and BD showed increased ALP activity compared to those cultured with control and GIC. Especially, BD loaded with 2% nBG induced the highest ALP activity, indicating that this nanoparticle content favors the stimulation of the cell differentiation process towards a mineralizing lineage. These results can be explained by the demonstrated capacity of nBGs to induce both osteogenic [34,67] and odontogenic differentiation [68–70]. nBGs have shown their capacity to promote the migration, adhesion and expression of odontogenic-related proteins and genes in hDPSCs, which has been mainly attributed to the release of Si and Ca ions [53,68,70]. In contrast with microsized particles, nBGs generate a faster release of soluble ions, which are capable of chemically driving hDPSCs along a mineralization pathway [68]. In addition, the application of microsized BG in a pulp capping procedure in primary human teeth has been shown to promote dentin bridge formation [58].

hDPSCs were used in this study to evaluate the cytocompatibility and bioactivity of the nanocomposite developed, since they are widely used in in vitro studies for the evaluation of dental materials. Their isolation is not invasive (using extracted semi-included third molars) and allows the creation of an approximation of how the cells of the dental pulp will behave when they come into direct contact with the dental material and its components [71]. One of the relevant characteristics of hDPSCs is their differentiation potential, which can be into odontoblasts, osteocytes/osteoblasts, adipocytes, chondrocytes and neural cells [72]. This capacity for differentiation allows the dental pulp to form dentin in response to a stimulus such as dental caries or trauma injuries. However, for this to occur, the vitality of the dental pulp must be preserved, hence the importance of evaluating cytocompatibility in materials for VPT [73].

Regarding the limitations of this study, additional studies are necessary to establish the safety and efficacy of the use of this material for clinical application. The results provide information about the in vitro cellular responses and whether these responses will be replicated in clinical conditions remains unknown. In addition, although the ALP activity of hDPSCs was analyzed to explore possible cell differentiation into a mineralizing type, the gene expression related to the formation of dentin was not studied. During dentinogenesis, several other genes, proteins and markers are detected, such as DSPP, dentin sialoprotein (DSP), dentin phosphoprotein (DPP), dentine matrix protein-1 (DMP-1) and OC, whose expression would be relevant to study [73]. nBGs have been shown to increase the expression of odontogenic genes, osteocalcin and DSPP protein production in hDPSCs [49]. Therefore, it would be interesting to know if this effect is maintained when they are incorporated into BD. Further investigations are also necessary to study the inflammatory response to this material, since inflammation and regeneration are of particular significance within the non-extensive space in the dental pulp tissue [69,74]. It would also be relevant to perform future research using animal models, which could provide histological evidence of hard tissue deposition and sealing ability, before clinical testing [75,76].

Within the limitations of this in vitro study, the results indicate that this nanocomposite could be a promising material for use in direct contact with the injured dentin–pulp complex, which could lead to faster repair. These outcomes raise the interesting possibility of using the nanocomposite in direct contact with dental pulp tissues to contribute to the preservation of natural dental tissue.

5. Conclusions

The incorporation of nBGs in a calcium silicate cement does not alter the cytocompatibility, in terms of viability, adhesion and morphology of hDPSCs, compared to the neat cement.

The nBG/BD nanocomposite exhibited a higher capacity to stimulate the differentiation of hDPSCs into a mineralizing lineage than BD.

The cellular properties of the nBG/BD nanocomposite make it a promising material to be used in VPT, which could lead to faster dentin formation and therefore to the healing and repair of the dentin–pulp complex.

Author Contributions: Conceptualization and design, C.C.N. and C.C.; performed the experiments, D.A.G. and M.M.; data analysis, D.A.G., J.M. and C.C.N.; wrote the manuscript, C.C.N., D.A.G. and C.C.; revised the manuscript, M.M. and J.M. All authors have read and agreed to the published version of the manuscript.

Funding: This research received no external funding.

Institutional Review Board Statement: The study was conducted according to the guidelines of the Declaration of Helsinki, and approved by the Ethics Committee of the Faculty of Dentistry, University of Chile (protocol code PRI-ODO2018/06).

Informed Consent Statement: Informed consent was obtained from all subjects involved in the study.

Data Availability Statement: The data presented in this study are available on request from the corresponding author.

Acknowledgments: The authors thank project FONDEQUIP EQM130076 and Rocio Orellana for her assistance in SEM sample preparation and imaging. Additionally, special thanks to Nicolas Menoni and Vittorio Basso, on behalf of Septodont, for providing us with Biodentine for this study. Thanks to Juan Fernández de los Ríos, from the Language and Translation Services, Direction of Academic Affairs, Faculty of Dentistry, Universidad de Chile, for kindly proofreading and checking the spelling and grammar of this article.

Conflicts of Interest: The authors declare no conflict of interest.

References

1. Aguilar, P.; Linsuwanont, P. Vital pulp therapy in vital permanent teeth with cariously exposed pulp: A systematic re-view. *J. Endod.* **2011**, *37*, 581–587. [CrossRef] [PubMed]
2. Elmsmari, F.; Ruiz, X.-F.; Miró, Q.; Feijoo-Pato, N.; Durán-Sindreu, F.; Olivieri, J.G. Outcome of Partial Pulpotomy in Cariously Exposed Posterior Permanent Teeth: A Systematic Review and Meta-analysis. *J. Endod.* **2019**, *45*, 1296–1306.e3. [CrossRef] [PubMed]
3. Wang, G.; Wang, C.; Qin, M. Pulp prognosis following conservative pulp treatment in teeth with complicated crown fractures-A retrospective study. *Dent. Traumatol.* **2017**, *33*, 255–260. [CrossRef] [PubMed]
4. Bimstein, E.; Rotstein, I. Cvek pulpotomy—Revisited. *Dent. Traumatol.* **2016**, *32*, 438–442. [CrossRef]
5. Bourguignon, C.; Cohenca, N.; Lauridsen, E.; Flores, M.T.; O'Connell, A.C.; Day, P.F.; Tsilingaridis, G.; Abbott, P.V.; Fouad, A.F.; Hicks, L.; et al. International Association of Dental Traumatology guidelines for the management of traumatic dental injuries: 1. Fractures and luxations. *Dent. Traumatol.* **2020**, *36*, 314–330. [CrossRef] [PubMed]
6. Duncan, H.F.; Galler, K.M.; Tomson, P.L.; Simon, S.; El-Karim, I.; Kundzina, R.; Krastl, G.; Dammaschke, T.; Fransson, H.; Markvart, M.; et al. European Society of Endodontology position statement: Management of deep caries and the exposed pulp. *Int. Endod. J.* **2019**, *52*, 923–934. [CrossRef]
7. Bakland, L.K.; Andreasen, J.O. Will mineral trioxide aggregate replace calcium hydroxide in treating pulpal and periodontal healing complications subsequent to dental trauma? A review. *Dent. Traumatol.* **2011**, *28*, 25–32. [CrossRef]
8. Kunert, M.; Lukomska-Szymanska, M. Bio-Inductive Materials in Direct and Indirect Pulp Capping—A Review Article. *Mater.* **2020**, *13*, 1204. [CrossRef]
9. Komabayashi, T.; Zhu, Q.; Eberhart, R.; Imai, Y. Current status of direct pulp-capping materials for permanent teeth. *Dent. Mater. J.* **2016**, *35*, 1–12. [CrossRef]
10. Schwendicke, F.; Brouwer, F.; Schwendicke, A.; Paris, S. Different materials for direct pulp capping: Systematic review and meta-analysis and trial sequential analysis. *Clin. Oral Investig.* **2016**, *20*, 1121–1132. [CrossRef]
11. Parirokh, M.; Torabinejad, M.; Dummer, P.M.H. Mineral trioxide aggregate and other bioactive endodontic cements: An updated overview—part I: Vital pulp therapy. *Int. Endod. J.* **2017**, *51*, 177–205. [CrossRef] [PubMed]
12. Asgary, S.; Eghbal, M.J.; Parirokh, M.; Ghanavati, F.; Rahimi, H. A comparative study of histologic response to different pulp capping materials and a novel endodontic cement. *Oral Surg. Oral Med. Oral Pathol. Oral Radiol. Endodontol.* **2008**, *106*, 609–614. [CrossRef] [PubMed]
13. Camilleri, J.; Montesin, F.E.; Brady, K.; Sweeney, R.; Curtis, R.V.; Ford, T.R.P. The constitution of mineral trioxide aggregate. *Dent. Mater.* **2005**, *21*, 297–303. [CrossRef] [PubMed]
14. Torabinejad, M.; White, D. Tooth Filling Material and Use. U.S. Patent 5,769,638, 23 June 1998.
15. Matsuura, T.; Ziauddin, S.M.; Kawata-Matsuura, V.K.S.; Sugimoto, K.; Yamada, S.; Yoshimura, A. Long-term clinical and radiographic evaluation of the effectiveness of direct pulp capping materials: A meta-analysis. *Dent. Mater. J.* **2021**, *40*, 1–7. [CrossRef]
16. Cushley, S.; Duncan, H.F.; Lappin, M.J.; Chua, P.; Elamin, A.D.; Clarke, M.; El-Karim, I.A. Efficacy of direct pulp cap-ping for management of cariously exposed pulps in permanent teeth: A systematic review and meta-analysis. *Int. Endod. J.* **2020**, *54*, 556–571. [CrossRef]
17. Asgary, S.; Shahabi, S.; Jafarzadeh, T.; Amini, S.; Kheirieh, S. The Properties of a New Endodontic Material. *J. Endod.* **2008**, *34*, 990–993. [CrossRef]
18. Yun, D.-A.; Park, S.-J.; Lee, S.-R.; Min, K.-S. Tooth discoloration induced by calcium-silicate-based pulp-capping materials. *Eur. J. Dent.* **2015**, *9*, 165–170. [CrossRef]
19. Bahabri, R.; Krsoum, M. Biodentine: Perforation, retrograde filling, and vital pulp therapy. A review. *Int. J. Med. Dent.* **2020**, *24*, 1–4.

20. Grech, L.; Mallia, B.; Camilleri, J. Investigation of the physical properties of tricalcium silicate cement-based root-end filling materials. *Dent. Mater.* **2013**, *29*, e20–e28. [CrossRef]
21. Nielsen, M.J.; Casey, J.A.; VanderWeele, R.A.; Vandewalle, K.S. Mechanical properties of new dental pulp-capping materials. *Gen. Dent.* **2016**, *64*, 44–48.
22. Corral, C.; Negrete, P.; Estay, J.; Osorio, S.; Covarrubias, C.; Junior, O.B.D.O.; Barud, H. Radiopacity and Chemical Assessment of New Commercial Calcium Silicate-Based Cements. *Int. J. Odontostomatol.* **2018**, *12*, 262–268. [CrossRef]
23. Brizuela, C.; Ormeño, A.; Cabrera, C.; Cabezas, R.; Silva, C.I.; Ramírez, V.; Mercade, M. Direct Pulp Capping with Calcium Hydroxide, Mineral Trioxide Aggregate, and Biodentine in Permanent Young Teeth with Caries: A Randomized Clinical Trial. *J. Endod.* **2017**, *43*, 1776–1780. [CrossRef] [PubMed]
24. Nowicka, A.; Lipski, M.; Parafiniuk, M.; Sporniak-Tutak, K.; Lichota, D.; Kosierkiewicz, A.; Kaczmarek, W.; Buczkowska-Radlińska, J. Response of Human Dental Pulp Capped with Biodentine and Mineral Trioxide Aggregate. *J. Endod.* **2013**, *39*, 743–747. [CrossRef]
25. Katge, F.A.; Patil, D.P. Comparative Analysis of 2 Calcium Silicate–based Cements (Biodentine and Mineral Trioxide Aggregate) as Direct Pulp-capping Agent in Young Permanent Molars: A Split Mouth Study. *J. Endod.* **2017**, *43*, 507–513. [CrossRef]
26. Harms, C.S.; Schäfer, E.; Dammaschke, T. Clinical evaluation of direct pulp capping using a calcium silicate cement—treatment outcomes over an average period of 2.3 years. *Clin. Oral Investig.* **2019**, *23*, 3491–3499. [CrossRef] [PubMed]
27. Sequeira, D.B.; Oliveira, A.R.; Seabra, C.M.; Palma, P.J.; Ramos, C.; Figueiredo, M.H.; Santos, A.C.; Cardoso, A.L.; Peça, J.; Santos, J.M. Regeneration of pulp-dentin complex using human stem cells of the apical papilla: In vivo interaction with two bioactive materials. *Clin. Oral Investig.* **2021**, 1–13. [CrossRef]
28. Luo, Z.; Kohli, M.R.; Yu, Q.; Kim, S.; Qu, T.; He, W.-X. Biodentine Induces Human Dental Pulp Stem Cell Differentiation through Mitogen-activated Protein Kinase and Calcium-/Calmodulin-dependent Protein Kinase II Pathways. *J. Endod.* **2014**, *40*, 937–942. [CrossRef]
29. Hanafy, A.K.; Shinaishin, S.F.; Eldeen, G.N.; Aly, R.M. Nano Hydroxyapatite & Mineral Trioxide Aggregate Efficiently Promote Odontogenic Differentiation of Dental Pulp Stem Cells. *Open Access Maced. J. Med. Sci.* **2018**, *6*, 1727–1731. [CrossRef] [PubMed]
30. Nam, S.; Won, J.-E.; Kim, C.-H.; Kim, H.-W. Odontogenic Differentiation of Human Dental Pulp Stem Cells Stimulated by the Calcium Phosphate Porous Granules. *J. Tissue Eng.* **2011**, *2*, 812547. [CrossRef] [PubMed]
31. Gu, Y.; Bai, Y.; Zhang, D. Osteogenic stimulation of human dental pulp stem cells with a novel gelatin-hydroxyapatite-tricalcium phosphate scaffold. *J. Biomed. Mater. Res. Part A* **2018**, *106*, 1851–1861. [CrossRef]
32. Catauro, M.; Barrino, F.; Poggetto, G.D.; Milazzo, M.; Blanco, I.; Ciprioti, S.V. Structure, drug absorption, bioactive and antibacterial properties of sol-gel SiO2/ZrO2 materials. *Ceram. Int.* **2020**, *46*, 29459–29465. [CrossRef]
33. Jones, J.R. Review of bioactive glass: From Hench to hybrids. *Acta Biomater.* **2013**, *9*, 4457–4486. [CrossRef] [PubMed]
34. Hench, L.L. The story of Bioglass®. *J. Mater. Sci. Mater. Med.* **2006**, *17*, 967–978. [CrossRef]
35. Hench, L.L. Bioceramics: From Concept to Clinic. *J. Am. Ceram. Soc.* **1991**, *74*, 1487–1510. [CrossRef]
36. Rodríguez-Lozano, F.; Collado-González, M.; Tomás-Catalá, C.; García-Bernal, D.; López, S.; Oñate-Sánchez, R.; Moraleda, J.; Murcia, L. GuttaFlow Bioseal promotes spontaneous differentiation of human periodontal ligament stem cells into cementoblast-like cells. *Dent. Mater.* **2019**, *35*, 114–124. [CrossRef]
37. Santos, J.M.; Pereira, S.; Sequeira, D.B.; Messias, A.L.; Martins, J.B.; Cunha, H.; Palma, P.J.; Santos, A.C. Biocompatibility of a bioceramic silicone-based sealer in subcutaneous tissue. *J. Oral Sci.* **2019**, *61*, 171–177. [CrossRef] [PubMed]
38. Abuna, G.; Campos, P.; Hirashi, N.; Giannini, M.; Nikaido, T.; Tagami, J.; Sinhoreti, M.A.C.; Geraldeli, S. The ability of a nanobioglass-doped self-etching adhesive to re-mineralize and bond to artificially demineralized dentin. *Dent. Mater.* **2021**, *37*, 120–130. [CrossRef]
39. Carvalho, E.M.; Ferreira, P.V.C.; Gutiérrez, M.F.; Sampaio, R.F.; Carvalho, C.N.; de Menezes, A.S.; Loguercio, A.D.; Bauer, J. Development and characterization of self-etching adhesives doped with 45S5 and niobophosphate bioactive glasses: Physicochemical, mechanical, bioactivity and interface properties. *Dent. Mater.* **2021**, *10*, 0109–5641. [CrossRef]
40. Jang, J.-H.; Lee, M.G.; Ferracane, J.L.; Davis, H.; Bae, H.E.; Choi, N.; Kim, D.-S. Effect of bioactive glass-containing resin composite on dentin remineralization. *J. Dent.* **2018**, *75*, 58–64. [CrossRef]
41. Korkut, E.; Torlak, E.; Altunsoy, M. Antimicrobial and mechanical properties of dental resin composite containing bioactive glass. *J. Appl. Biomater. Funct. Mater.* **2016**, *14*, e296–e301. [CrossRef] [PubMed]
42. Khvostenko, D.; Hilton, T.; Ferracane, J.; Mitchell, J.; Kruzic, J. Bioactive glass fillers reduce bacterial penetration into marginal gaps for composite restorations. *Dent. Mater.* **2016**, *32*, 73–81. [CrossRef]
43. Par, M.; Spanovic, N.; Tauböck, T.T.; Attin, T.; Tarle, Z. Degree of conversion of experimental resin composites containing bioactive glass 45S5: The effect of post-cure heating. *Sci. Rep.* **2019**, *9*, 17245. [CrossRef]
44. Tiskaya, M.; Shahid, S.; Gillam, D.; Hill, R. The use of bioactive glass (BAG) in dental composites: A critical review. *Dent. Mater.* **2021**, *37*, 296–310. [CrossRef]
45. Yang, S.-Y.; Piao, Y.-Z.; Kim, S.-M.; Lee, Y.-K.; Kim, K.-N.; Kim, K.-M. Acid neutralizing, mechanical and physical properties of pit and fissure sealants containing melt-derived 45S5 bioactive glass. *Dent. Mater.* **2013**, *29*, 1228–1235. [CrossRef] [PubMed]
46. Khvostenko, D.; Mitchell, J.C.; Hilton, T.J.; Ferracane, J.L.; Kruzic, J.J. Mechanical performance of novel bioactive glass containing dental restorative composites. *Dent. Mater.* **2013**, *29*, 1139–1148. [CrossRef]

47. Khodaei, M.; Nejatidanesh, F.; Shirani, M.J.; Valanezhad, A.; Watanabe, I.; Savabi, O. The effect of the nano- bioglass reinforcement on magnesium based composite. *J. Mech. Behav. Biomed. Mater.* **2019**, *100*, 103396. [CrossRef]
48. Misra, S.K.; Mohn, D.; Brunner, T.J.; Stark, W.J.; Philip, S.E.; Roy, I.; Salih, V.; Knowles, J.C.; Boccaccini, A.R. Comparison of nanoscale and microscale bioactive glass on the properties of P(3HB)/Bioglass® composites. *Biomaterials* **2008**, *29*, 1750–1761. [CrossRef] [PubMed]
49. Ajita, J.; Saravanan, S.; Selvamurugan, N. Effect of size of bioactive glass nanoparticles on mesenchymal stem cell proliferation for dental and orthopedic applications. *Mater. Sci. Eng. C* **2015**, *53*, 142–149. [CrossRef] [PubMed]
50. Vollenweider, M.; Brunner, T.J.; Knecht, S.; Grass, R.N.; Zehnder, M.; Imfeld, T.; Stark, W.J. Remineralization of human dentin using ultrafine bioactive glass particles. *Acta Biomater.* **2007**, *3*, 936–943. [CrossRef] [PubMed]
51. Waltimo, T.; Brunner, T.; Vollenweider, M.; Stark, W.; Zehnder, M. Antimicrobial Effect of Nanometric Bioactive Glass 45S5. *J. Dent. Res.* **2007**, *86*, 754–757. [CrossRef]
52. Odermatt, R.; Par, M.; Mohn, D.; Wiedemeier, D.B.; Attin, T.; Tauböck, T.T. Bioactivity and Physico-Chemical Properties of Dental Composites Functionalized with Nano- vs. Micro-Sized Bioactive Glass. *J. Clin. Med.* **2020**, *9*, 772. [CrossRef] [PubMed]
53. Gong, W.; Huang, Z.; Dong, Y.; Gan, Y.; Li, S.; Gao, X.; Chen, X. Ionic Extraction of a Novel Nano-sized Bioactive Glass Enhances Differentiation and Mineralization of Human Dental Pulp Cells. *J. Endod.* **2014**, *40*, 83–88. [CrossRef]
54. Lee, J.-H.; Kang, M.-S.; Mahapatra, C.; Kim, H.-W. Effect of Aminated Mesoporous Bioactive Glass Nanoparticles on the Differentiation of Dental Pulp Stem Cells. *PLoS ONE* **2016**, *11*, e0150727. [CrossRef] [PubMed]
55. Oguntebi, B.; Clark, A.; Wilson, J. Pulp Capping with Bioglass® and Autologous Demineralized Dentin in Miniature Swine. *J. Dent. Res.* **1993**, *72*, 484–489. [CrossRef] [PubMed]
56. Long, Y.; Liu, S.; Zhu, L.; Liang, Q.; Chen, X.; Dong, Y. Evaluation of Pulp Response to Novel Bioactive Glass Pulp Capping Materials. *J. Endod.* **2017**, *43*, 1647–1650. [CrossRef]
57. Zhu, N.; Chatzistavrou, X.; Papagerakis, P.; Ge, L.; Qin, M.; Wang, Y. Silver-Doped Bioactive Glass/Chitosan Hydrogel with Potential Application in Dental Pulp Repair. *ACS Biomater. Sci. Eng.* **2019**, *5*, 4624–4633. [CrossRef] [PubMed]
58. Ahmadvand, M.; Haghgoo, R. Evaluation of pulpal response of deciduous teeth after direct pulp capping with bioactive glass and mineral trioxide aggregate. *Contemp. Clin. Dent.* **2016**, *7*, 332–335. [CrossRef] [PubMed]
59. Hanna, S.N.; Alfayate, R.P.; Prichard, J. Vital pulp therapy an insight over the available literature and future expectations. *Eur. Endod. J.* **2020**, *5*, 46–53.
60. Nuñez, C.C.; Covarrubias, C.; Fernandez, E.; De Oliveira, O.B. Enhanced bioactive properties of BiodentineTM modified with bioactive glass nanoparticles. *J. Appl. Oral Sci.* **2017**, *25*, 177–185. [CrossRef] [PubMed]
61. Valenzuela, F.; Covarrubias, C.; Martínez, C.; Smith, P.; Díaz-Dosque, M.; Yazdani-Pedram, M. Preparation and bioactive properties of novel bone-repair bionanocomposites based on hydroxyapatite and bioactive glass nanoparticles. *J. Biomed. Mater. Res. Part B Appl. Biomater.* **2012**, *100*, 1672–1682. [CrossRef]
62. Covarrubias, C.; Agüero, A.; Maureira, M.; Morelli, E.; Escobar, G.; Cuadra, F.; Peñafiel, C.; Von Marttens, A. In situ preparation and osteogenic properties of bionanocomposite scaffolds based on aliphatic polyurethane and bioactive glass nanoparticles. *Mater. Sci. Eng. C* **2019**, *96*, 642–653. [CrossRef]
63. Sevari, S.P.; Shahnazi, F.; Chen, C.; Mitchell, J.C.; Ansari, S.; Moshaverinia, A. Bioactive glass-containing hydrogel delivery system for osteogenic differentiation of human dental pulp stem cells. *J. Biomed. Mater. Res. Part A* **2019**, *108*, 557–564. [CrossRef]
64. Kim, G.-H.; Park, Y.-D.; Lee, S.-Y.; El-Fiqi, A.; Kim, J.-J.; Lee, E.-J.; Kim, H.-W.; Kim, E.-C. Odontogenic stimulation of human dental pulp cells with bioactive nanocomposite fiber. *J. Biomater. Appl.* **2014**, *29*, 854–866. [CrossRef] [PubMed]
65. Zhu, N.; Chatzistavrou, X.; Ge, L.; Qin, M.; Papagerakis, P.; Wang, Y. Biological properties of modified bioactive glass on dental pulp cells. *J. Dent.* **2019**, *83*, 18–26. [CrossRef] [PubMed]
66. Nuñez, C.M.C.; Bosomworth, H.J.; Field, C.; Whitworth, J.M.; Valentine, R.A. Biodentine and Mineral Trioxide Aggregate Induce Similar Cellular Responses in a Fibroblast Cell Line. *J. Endod.* **2014**, *40*, 406–411. [CrossRef]
67. Wang, S.; Huang, G.; Dong, Y. Directional Migration and Odontogenic Differentiation of Bone Marrow Stem Cells Induced by Dentin Coated with Nanobioactive Glass. *J. Endod.* **2020**, *46*, 216–223. [CrossRef]
68. Wang, S.; Gao, X.; Gong, W.; Zhang, Z.; Chen, X.; Dong, Y. Odontogenic differentiation and dentin formation of dental pulp cells under nanobioactive glass induction. *Acta Biomater.* **2014**, *10*, 2792–2803. [CrossRef] [PubMed]
69. Mocquot, C.; Attik, N.; Pradelle-Plasse, N.; Grosgogeat, B.; Colon, P. Bioactivity assessment of bioactive glasses for dental applications: A critical review. *Dent. Mater.* **2020**, *36*, 1116–1143. [CrossRef]
70. Wang, S.; Hu, Q.; Gao, X.; Dong, Y. Characteristics and Effects on Dental Pulp Cells of a Polycaprolactone/Submicron Bioactive Glass Composite Scaffold. *J. Endod.* **2016**, *42*, 1070–1075. [CrossRef] [PubMed]
71. Morsczeck, C.; Reichert, T.E. Dental stem cells in tooth regeneration and repair in the future. *Expert Opin. Biol. Ther.* **2018**, *18*, 187–196. [CrossRef] [PubMed]
72. Nuti, N.; Corallo, C.; Chan, B.M.F.; Ferrari, M.; Gerami-Naini, B. Multipotent Differentiation of Human Dental Pulp Stem Cells: A Literature Review. *Stem Cell Rev. Rep.* **2016**, *12*, 511–523. [CrossRef] [PubMed]
73. Da Rosa, W.L.O.; Piva, E.; Da Silva, A.F. Disclosing the physiology of pulp tissue for vital pulp therapy. *Int. Endod. J.* **2018**, *51*, 829–846. [CrossRef]
74. Giraud, T.; Jeanneau, C.; Rombouts, C.; Bakhtiar, H.; Laurent, P.; About, I. Pulp capping materials modulate the balance between inflammation and regeneration. *Dent. Mater.* **2019**, *35*, 24–35. [CrossRef] [PubMed]

75. Song, M.; Kim, S.; Kim, T.; Park, S.; Shin, K.-H.; Kang, M.; Park, N.-H.; Kim, R. Development of a Direct Pulp-capping Model for the Evaluation of Pulpal Wound Healing and Reparative Dentin Formation in Mice. *J. Vis. Exp.* **2017**, *54973*, e54973. [CrossRef] [PubMed]
76. Dammaschke, T. Rat molar teeth as a study model for direct pulp capping research in dentistry. *Lab. Anim.* **2010**, *44*, 1–6. [CrossRef]

Article

The Effects of Three Remineralizing Agents on the Microhardness and Chemical Composition of Demineralized Enamel

Ivan Salinovic [1,*], Zdravko Schauperl [2], Marijan Marcius [3] and Ivana Miletic [1]

1. Department of Endodontics and Restorative Dentistry, School of Dental Medicine, University of Zagreb, 10000 Zagreb, Croatia; miletic@sfzg.hr
2. Department of Materials, Faculty of Mechanical Engineering and Naval Architecture, University of Zagreb, 10000 Zagreb, Croatia; zdravko.schauperl@fsb.hr
3. Division of Materials Chemistry, Ruđer Bošković Institute, 10000 Zagreb, Croatia; mmarcius@irb.hr
* Correspondence: isalinovic@sfzg.hr

Abstract: This study aimed to determine the effects of three different varnish materials (containing casein phosphopeptide-amorphous calcium phosphate, nano-hydroxyapatite, and fluoride) on enamel. Thirty-three extracted human third molars were used for specimen preparation. These were demineralized using phosphoric acid. Three experimental groups (n = 11) were treated with 3M™ Clinpro™ White Varnish, MI Varnish®, and Megasonex® toothpaste, respectively, every twenty-four hours for fourteen days. Analysis of the microhardness of the specimens' enamel surfaces was carried out via the Vickers method, and by scanning electron microscopy/energy dispersive X-ray spectroscopy (SEM/EDS). Analysis was performed at three stages: at baseline value, after demineralization, and after the period of remineralization. Data were subjected to Scheffe's post hoc test. The mean microhardness values (HV0.1) obtained for the group of samples treated with MI Varnish® were higher compared with the other two groups ($p = 0.001$ for both comparisons), while the first and third groups did not differ significantly from each other ($p = 0.97$). SEM analysis showed uneven patterns and porosities on all samples tested. EDS results showed an increase in the mineral content of the examined samples, with the highest mineral content observed in the MI Varnish® group. It can be concluded that MI Varnish® use has a better remineralization effect on enamel than the other two materials.

Keywords: remineralizing dental materials; hydroxyapatite; calcium phosphates; ion release; varnish materials; enamel remineralization; bioactive coatings

1. Introduction

Enamel is the outermost layer of the tooth; therefore, it is susceptible to various influences and conditions of the oral cavity. Enamel is primarily composed of hydroxyapatite ($Ca_{10}(PO_4)_6(OH)_2$) [1], which is a crystalline form of calcium phosphate. While it is the hardest tissue in the human body [2], it is highly prone to demineralization, which occurs in acidic environments and is usually accelerated by the consumption of acidic and sugar-containing foods [3]. On the other hand, antibacterial agents, saliva, and ions (e.g., fluoride, calcium, and phosphate) promote remineralization. The balance between the continual remineralization and demineralization processes is extremely fragile; thus, when disturbed for a long period, imbalance usually leads to early carious lesions [4].

The minimally invasive approach in dentistry dictates the usage of treatments that avoid the loss of hard dental tissues [5,6]. One such treatment involves the remineralization of early, noncavitated lesions, as opposed to performing conventional treatment [7]. To date, fluoride is considered the principal remedy for halting mineral loss from enamel; although, materials which can serve as reservoirs and release fluoride continuously have been developed [8,9]. The introduction of fluoride for preventative uses in dentistry was revolutionary, resulting in a substantial decline in caries prevalence [10,11]. Aside from its affinity for treatment of hard dental tissues [12], fluoride leads to the formation of fluorapatite. Fluorapatite is less susceptible to acidic effects than hydroxyapatite, as it replaces hydroxyl groups in enamel apatite [13,14]. However, the downside of fluoride usage is that it usually leads to superficial remineralization, leaving the deeper layers unaffected [15].

However, new findings on remineralization materials and mechanisms have led to the development of new courses of action concerning the promotion of demineralized enamel regeneration. The introduction of casein phosphopeptide-amorphous calcium phosphate (CPP-ACP) as a remineralizing agent marked a major breakthrough in arresting early carious lesions [16]. The usage of CPP-ACP makes the stabilization of calcium phosphate on dental surfaces and in dental plaques possible. CPP-ACP creates non-cariogenic plaques that release calcium and phosphate ions, promoting remineralization in acidic environments [17]. These findings have led to the release of varnishes containing both CPP-ACP and high amounts of fluorides.

Nano-hydroxyapatite, which resembles the structure of apatite crystals found in hard dental tissues, can replace the natural mineral content of enamel [18,19]. Its advantages are primarily due to its size, as nanosized particles dissolve better, enabling a faster reaction. Additionally, particles of that size fit better into enamel-induced defects; thus, creating a more compact surface and preventing further demineralization [20,21]. However, there is still no proof that any of the nano-hydroxyapatite-containing products are more effective than fluorides [22,23].

Enamel remineralization can be observed by testing its microhardness and conducting scanning electron microscopy in conjunction with energy dispersive X-ray spectroscopy (SEM/EDS), both of which are useful for observing the surface of enamel and determining its mineral composition [24,25].

The rising number of commercially available products for enamel remineralization can create confusion for operators, as there have not been enough studies conducted to suggest the most appropriate one. As such, this study aims to determine the effects of three widely used varnish materials which are labeled as bioactive (containing fluorides, CPP-ACP, and nano-hydroxyapatite) on enamel remineralization.

The null hypotheses are:
1. There will be no difference in the effect on enamel microhardness between the tested materials.
2. There will be no difference in the mineral composition of the specimens treated with the tested materials.
3. The micro-surface of all specimens will be the same.

2. Materials and Methods

The protocol for the current study was approved by the Ethics Committee of the School of Dental Medicine, University of Zagreb (No. 05-PA-30-XXVII-5/2021), Zagreb, Croatia. In total, 33 healthy human third molars were collected at the Department of Oral Surgery, School of Dental Medicine, University of Zagreb, Zagreb, Croatia.

2.1. Sample Preparation

Prior to use, the teeth were thoroughly cleaned using a scaler and brushes and were stored in saline solution at room temperature. The tested materials are listed in Table 1.

Table 1. List of the materials used in the study and their respective active ingredients.

Material	Active Ingredients	Manufacturer
3M™ Clinpro™ White Varnish	22,600 ppm fluoride	3M ESPE, St. Paul, MN, USA
MI Varnish®	Casein phosphopeptide-amorphous calcium phosphate (CPP-ACP), 5% sodium fluoride	GC Corporation, Tokyo, Japan
Megasonex®	Nano-hydroxyapatite	Panaford B.V., Rotterdam, The Netherlands

The teeth were embedded in autopolymerizing acrylic resin (Heraus Kuzler Gmbh, Hanau, Germany) and left to set for 24 h, forming rectangular blocks. The samples were then polished using a Mintech 233 (Presi, Le Loche, Switzerland) polishing machine at a speed of 300 RPM, with water cooling. Standard metallographic grinding paper was used, from rough to fine (P320, P600, P1200, and P2400), exposing the smooth enamel surface, in order to perform optimal microhardness testing and SEM/EDS analysis. The samples were randomly divided into three groups for the evaluation of microhardness (n = 10). One additional specimen for each group was used in order to make samples for SEM/EDS analysis. These were cut into thin slices of approximately 1.5 mm thickness using an IsoMet 1000 Precision Cutter (Buehler, Lake Bluff, IL, USA) and an IsoMet Diamond Wafering Blade with 12.7 mm arbor size and 0.5 mm thickness (Buehler, Lake Bluff, IL, USA), at a speed of 250 rounds per minute.

2.2. Demineralization and Remineralization Cycle

The exposed buccal enamel surface was demineralized using 37% phosphoric acid (DiaDent Group International, Chungcheongbuk-do, Korea) for three minutes. The samples were then washed and air-dried.

Three different remineralizing agents were then applied twice a day for two minutes, using a soft applicator brush (3M ESPE, St. Paul, IL, USA), and stored in saline (Croatian Institute of Transfusion Medicine, Zagreb, Croatia) at room temperature over a period of 14 days. Saline was freshly prepared every two days.

2.3. Vickers Microhardness Measurement

Microhardness of samples was determined using the Vickers microhardness tester KBW 1-V (KB Prüftechnik GmbH, Hochdorf-Assenheim, Germany) and was performed at the Department of Endodontics and Restorative Dentistry, School of Dental Medicine, University of Zagreb. This method is based on observing the enamel's resistance to plastic deformation. Microhardness was measured at three stages: baseline value, after demineralization, and after the remineralization period. The load used for the microhardness measurement was 100 g (HV0.1) and was applied for 10 seconds, as suggested by Farooq [26]. On each sample, three indents were made, and the mean value was calculated.

2.4. SEM/EDS Analysis

SEM/EDS analysis was performed on one specimen for each material tested at the Ruđer Bošković Institute, Division of Materials Chemistry, Zagreb, Croatia. The microscope model used was the JSM-7000F (JEOL Ltd., Tokyo, Japan), and the Inca 350 EDS System (Oxford Instruments, High Wycombe, UK) was used for EDS. Preceding the examinations, the samples were polished using a brush and air-dried. The surfaces of the samples, as well as the share of certain chemical elements, were observed.

2.5. Statistical Analysis

The results were analyzed using descriptive statistics (mean, standard deviation), and statistical inferences were made using a mixed-design ANOVA, which included repeated measures testing (for microhardness changes due to demineralization and remineralization) and independent sample analysis to compare differences across groups. The post hoc differences were calculated using Scheffe's post hoc test, for analysis, where the ANOVA results were significant. Since differences between microhardness were found in different sub-samples before intervention, the effects of initial hardness were assessed using the Pearson correlation coefficient.

3. Results

The results of microhardness testing are shown in Table 2. The baseline values for all three groups were statistically significantly different from each other (ANOVA test for three groups regarding initial microhardness values, $p = 0.000$). All differences are statistically significant, with $p = 0.024$ for the first group vs. the second group, $p = 0.008$ for the first group vs. the third group and 0.000 s vs. third group (Scheffe's post hoc test). The correlations between these variables are not statistically significant, which indicates that the level of hardness at the beginning does not affect the level of hardness at the end ($r = -0.301$, $p = 0.106$), nor does it affect the level of hardness after demineralization. This is further proven by the fact that there is no significant difference among the groups tested after demineralization (ANOVA test $p = 0.362$).

Table 2. Microhardness of the three different groups in three stages (HV 0.1).

Microhardness Groups	Baseline	After Demineralization	After Remineralization
3M™ Clinpro™ White Varnish	366.00 ± 18.93	190.30 ± 23.71	236.57 ± 19.41
MI Varnish®	343.52 ± 26.66	192.73 ± 16.37	286.65 ± 34.07
Megasonex®	393.05 ± 16.14	201.90 ± 15.30	237.97 ± 32.52

After remineralization, the differences between the groups were statistically significant (ANOVA test for three groups and the mean microhardness values after demineralization, $p = 0.000$)—in the sense that the difference was significant between the samples in the second group (MI Varnish®) compared with the other two (Scheffe's post hoc test, $p = 0.001$ for both comparisons)—while the first and third groups, 3M™ Clinpro™ White Varlish and Megasonex®, did not differ from each other (Scheffe's post hoc test, $p = 0.97$).

SEM analysis showed normal enamel surfaces prior to demineralization (Figure 1). Much more diverse surfaces were observed after remineralization (Figure 2), with uneven patterns, porosities, material deposits, and debris.

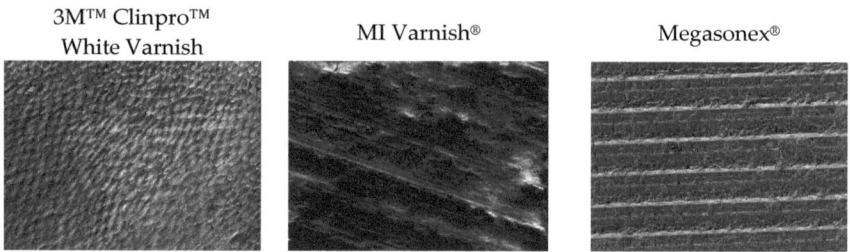

Figure 1. Representative SEM images (accelerating voltage of 5 kV, working distance of 10 mm, magnification of 500×) of the sample surfaces before remineralization.

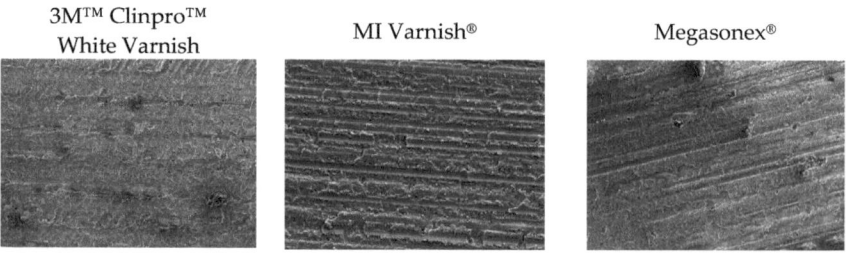

Figure 2. Representative SEM images (accelerating voltage of 5 kV, working distance of 10 mm, magnification of 500×) of the sample surfaces after remineralization.

Figure 3 shows the results of EDS analysis, before and after remineralization, with the share of particular elements varying greatly among the samples tested.

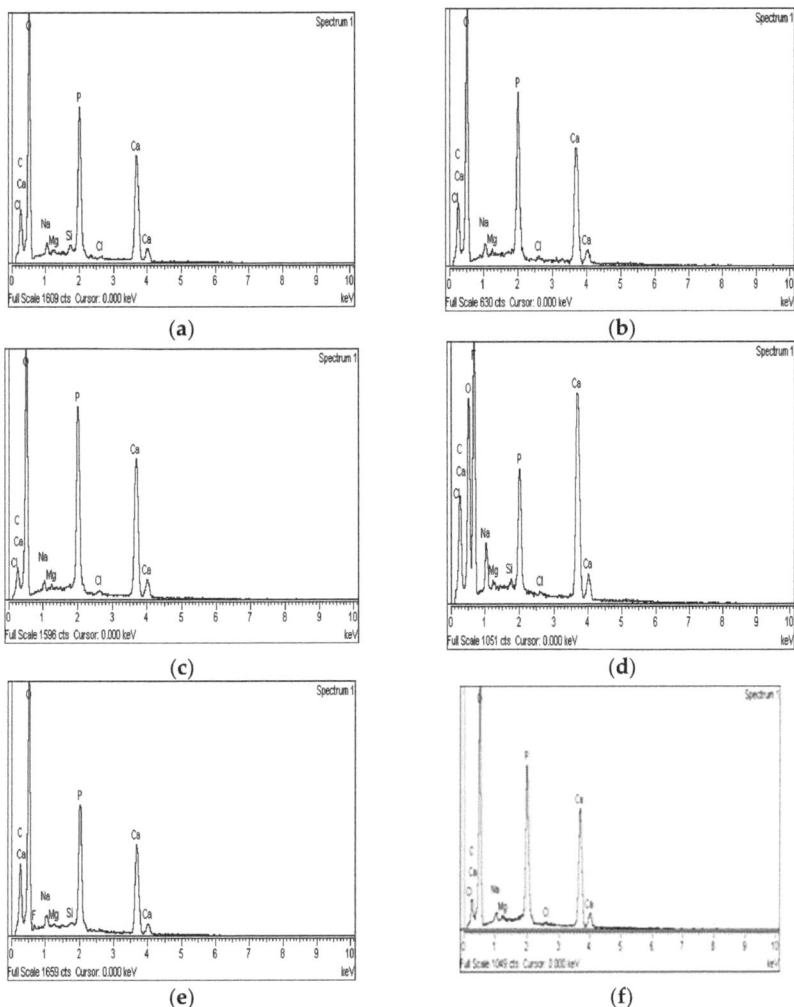

Figure 3. Representative results of EDS elemental analysis, following demineralization and remineralization of the samples treated with 3M™ Clinpro™ White Varnish (**a,b**), MI Varnish® (**c,d**) and Megasonex® (**e,f**).

4. Discussion

The present study evaluated the microhardness of enamel in order to determine the extent of remineralization. This method has also been used by other authors [27,28].

In order to create an artificial demineralized lesion, 37% phosphoric acid was used, as suggested by Sorozini [29]. Within the limitations of this study, this was considered sufficient, as it has been shown that absolute simulation of oral conditions is almost impossible due to other variables, including the speed of saliva flow and its buffering ability, dynamic pH cycles in the mouth, and behavioral changes [28,30,31]. As expected, the application of acid led to a significant decrease in microhardness values due to mineral loss, lowering the values in each group to similar levels. Regarding the duration of remineralization during the trial, multiple timespans have been used: from 24 h by Ali et al. [32] to 30 days by Balakrishnan [33]. In the present study, we opted for a 14-day duration to obtain more comparable results with previous studies, as it has been shown that the extent of remineralization increases over time [33,34].

The study results show that the usage of MI Varnish® leads to higher microhardness values compared with the other two materials; accordingly, the first null hypothesis was rejected.

While all three materials tested lead to certain microhardness increases, MI Varnish® stands out, possibly due to its high CPP-ACP and high fluoride content. CPP-ACP binds firmly to the enamel surface and subsequently arrests fluoride ions, keeping them closer to the enamel surface for a longer period [35]. The use of fluoride as a remineralizing agent is based on its formation of fluorapatite, which requires large amounts of calcium and phosphate ions. Therefore, a lack of those ions is possible when fluoride alone is applied topically [34]. This could explain why the microhardness values obtained for 3M™ Clinpro™ White Varnish, which contains only fluorides as an active ingredient, are lower. These findings were previously confirmed by Reynolds et al. [15]. As suggested by Wegehaupt et al. [36], an increase in microhardness in both cases is caused by the accumulation of fluoride-containing compounds on enamel.

Alternatively, the results from Vyavhare et al. [37] revealed that CPP-ACP did not show superior surface remineralization compared with fluorides, claiming that it mostly causes subsurface remineralization by allowing calcium and phosphate ions to penetrate deeper; thus, leading to remineralization throughout the entire lesion [38]. However, the present study included a CPP-ACP material, which also contained fluoride ions. The combination of both CPP-ACP and fluorides creates stabilized amorphous calcium fluoride phosphate, resulting in the increased incorporation of fluoride ions and increased amounts of calcium and phosphate ions on the enamel surface [34,39,40].

When it comes to nano-hydroxyapatite-containing Megasonex® toothpaste, the obtained microhardness values were similar to those of 3M™ Clinpro™ White Varnish, which contains only fluorides. Juntavee [41] and Amaechi [42] reported similar results. These findings suggest that nano-hydroxyapatite could be used as an alternative to products containing only fluoride [43]. This is important, as fluorides have recently been recognized as a possible neurotoxin [44]; therefore, increasing numbers of people are avoiding it [44]. However, further investigation of this topic is required.

Huang et al. [45] concluded that different concentrations of nano-hydroxyapatite were able to remineralize enamel at each time point in pH cycling; moreover, the optimal concentration of nano-hydroxyapatite for enamel remineralization proved to be a 10% suspension [46].

Another important factor to consider regarding the use of nano-hydroxyapatite is its particle size. A study by Huang et al. [45] showed that suspensions containing particles ranging from 10 nm to 20 nm in diameter only caused remineralization of the superficial layer in demineralized lesions. Since a previous study by Tschoppe et al. showed that nano-hydroxyapatite particles >20 nm enhance remineralization throughout the lesion [21], in the present study, the particle size was 20–50 nm in diameter.

As a method to gain insight into the changes in mineral content of the samples, SEM/EDS analysis was performed. SEM is useful for observing the changes in surface structure, whereas EDS is considered the gold standard for observing the loss and gain of minerals during the remineralization of an artificial demineralized lesion [47].

Since there were noteworthy differences in the surface microstructure and mineral content of the tested samples after remineralization, the second and third null hypotheses were partially rejected.

In order to avoid influencing the appearance of the samples' micro-surfaces and their mineral contents, they were polished using only a brush prior to SEM/EDS. Markings made by the saw during the cutting process are visible. Additionally, clusters of different minerals and possibly additives from the tested materials can be observed. A smearlike layer, which was probably created during sample preparation, was found on all tested specimens.

EDS analysis provided insight into the chemical changes that occurred during the demineralization and remineralization processes. Following demineralization, lower concentrations of calcium and phosphate were observed, confirming the expected mineral loss. However, compared with findings by Wang et al., [48] the values in the present study are significantly higher. This is because samples in the present study were demineralized using only phosphoric acid, without pH cycling, and because the examined surface was enamel, which contains more minerals than dentine.

In all three groups, increased calcium and phosphate were observed when analyzing the EDS results after demineralization: the highest was in the MI Varnish® group, where fluoride was also found. This is possibly due to the fact that CPP-ACP from the material arrested fluoride on the surface, which was then identified during analysis.

Within their limitations, such as the sample size and the lack of pH cycles, these types of in vitro studies provide insight into the effects that bioactive materials can produce, allowing an evaluation of their possible behavior in clinical settings; however, the results should be observed with a consideration of the experimental conditions. As such, further research is required to obtain more explanations and examples of their abilities.

5. Conclusions

MI Varnish®, a CPP-ACP and fluoride-based material, shows a higher capacity to remineralize enamel—as seen in its influence on the microhardness and chemical composition—compared with 3M™ Clinpro™ White Varnish and Megasonex® toothpaste, which contain only fluorides and nano-hydroxyapatite, respectively, as active ingredients. The joint effect of CPP-ACP and fluorides is a promising combination and a vital agent for the treatment of demineralized lesions.

Author Contributions: Conceptualization, I.S. and I.M.; methodology, I.S. and Z.S.; software, M.M.; validation, I.M., Z.S. and M.M.; formal analysis, I.M.; investigation, I.S.; resources, I.M.; data curation, I.M.; writing—original draft preparation, I.S.; writing—review and editing, I.S. and I.M; visualization, I.S.; supervision, I.M.; project administration, I.M.; funding acquisition, I.M. All authors have read and agreed to the published version of the manuscript.

Funding: The research was conducted within the Croatian Science Foundation project 'Research and Development of New Micro and Nanostructural Bioactive Materials in Dental Medicine', BIODENTMED No. IP-2018-01-1719.

Institutional Review Board Statement: The study was conducted according to the guidelines of the Declaration of Helsinki and approved by the Institutional Ethics Committee of University of Zagreb, School of Dental Medicine (No. 05-PA-30-XXVII-5/2021).

Informed Consent Statement: Informed consent was obtained from all subjects involved in the study.

Data Availability Statement: The data presented in this study are available on request from the corresponding author.

Conflicts of Interest: The authors declare no conflict of interest.

References

1. Wen, S.L. Human Enamel Structure Studied by High Resolution Electron Microscopy. *Electron. Microsc. Rev.* **1989**, *2*, 1–16. [PubMed]
2. Gil-Bona, A.; Bidlack, F.B. Tooth Enamel and its Dynamic Protein Matrix. *Int. J. Mol. Sci.* **2020**, *21*, 4458. [CrossRef] [PubMed]
3. Pinelli, M.-D.-M.; Catelan, A.; De Resende, L.-F.-M.; Soares, L.-E.-S.; Aguiar, F.H.B.; Liporoni, P.-C.-S. Chemical composition and roughness of enamel and composite after bleaching, acidic beverages and toothbrushing. *J. Clin. Exp. Dent.* **2019**, *11*, e1175–e1180. [CrossRef] [PubMed]
4. Tsai, M.-T.; Wang, Y.-L.; Yeh, T.-W.; Lee, H.-C.; Chen, W.-J.; Ke, J.-L.; Lee, Y.-J. Early detection of enamel demineralization by optical coherence tomography. *Sci. Rep.* **2019**, *9*, 1–9. [CrossRef] [PubMed]
5. Ericson, D.; Kidd, E.; McComb, D.; Mjör, I.; Noack, M.J. Minimally Invasive Dentistry—Concepts and techniques in cariology. *Oral Health Prev. Dent.* **2003**, *1*, 59–72. [PubMed]
6. Christensen, G.J. The advantages of minimally invasive dentistry. *J. Am. Dent. Assoc.* **2005**, *136*, 1563–1565. [CrossRef] [PubMed]
7. Villalobos-Rodelo, J.J.; Medina-Solís, C.E.; Verdugo-Barraza, L.; Islas-Granillo, H.; García-Jau, R.A.; Escoffié-Ramírez, M.; Maupomé, G. Experience of Non-Reversible and Reversible Carious Lesions in 11 and 12 Years Old Mexican Schoolchildren: A Negative Binomial Regression Analysis. *Biomedica* **2013**, *33*, 88–98. [PubMed]

8. Parkin, N.; Dyer, F.; Millett, D.T.; Furness, S.; Germain, P. Fluorides for the prevention of early tooth decay (demineralised white lesions) during fixed brace treatment. *Cochrane Database Syst. Rev.* **2013**, CD003809. [CrossRef]
9. Shahid, M. Regular supervised fluoride mouthrinse use by children and adolescents associated with caries reduction. *Evid.-Based Dent.* **2017**, *18*, 11–12. [CrossRef]
10. Fejerskov, O. Changing Paradigms in Concepts on Dental Caries: Consequences for Oral Health Care. *Caries Res.* **2004**, *38*, 182–191. [CrossRef]
11. Philip, N. State of the Art Enamel Remineralization Systems: The Next Frontier in Caries Management. *Caries Res.* **2019**, *53*, 284–295. [CrossRef]
12. National Research Council. *Health Effects of Ingested Fluoride*; National Academies Press: Washington, DC, USA, 1993.
13. Al-Noaman, A.; Karpukhina, N.; Rawlinson, S.; Hill, R.G. Effect of FA on bioactivity of bioactive glass coating for titanium dental implant. Part I: Composite powder. *J. Non-Cryst. Solids* **2013**, *364*, 92–98. [CrossRef]
14. Stanić, V.; Radosavljević-Mihajlović, A.S.; Živković-Radovanović, V.; Nastasijević, B.; Marinović-Cincović, M.; Marković, J.P.; Budimir, M.D. Synthesis, Structural Characterisation and Antibacterial Activity of Ag^+-Doped Fluorapatite Nanomaterials Prepared by Neutralization Method. *Appl. Surf. Sci.* **2015**, *337*, 72–80. [CrossRef]
15. Reynolds, E. Remineralization of Enamel Subsurface Lesions by Casein Phosphopeptide-stabilized Calcium Phosphate Solutions. *J. Dent. Res.* **1997**, *76*, 1587–1595. [CrossRef]
16. Kalra, D.; Kalra, R.; Kini, P.; Prabhu, C.A. Nonfluoride remineralization: An evidence-based review of contemporary technologies. *J. Dent. Allied Sci.* **2014**, *3*, 24. [CrossRef]
17. Naik, S.V.; Attiguppe, P.; Malik, N.; Ballal, S. CPP–ACP and Fluoride: A Synergism to Combat Caries. *Int. J. Clin. Pediatr. Dent.* **2019**, *12*, 120–125. [CrossRef]
18. Zhang, X.; Deng, X.; Wu, Y. Remineralising Nanomaterials for Minimally Invasive Dentistry. In *Nanotechnology in Endodontics: Current and Potential Clinical Applications*; Kishen, A., Ed.; Springer International Publishing: Cham, Switzerland, 2015; pp. 173–193.
19. Arifa, M.K.; Ephraim, R.; Rajamani, T. Recent Advances in Dental Hard Tissue Remineralization: A Review of Literature. *Int. J. Clin. Pediatr. Dent.* **2019**, *12*, 139–144. [CrossRef]
20. Amin, M.; Mehta, R.; Duseja, S.; Desai, K. Evaluation of the Efficacy of Commercially Available Nano-Hydroxyapatite Paste as a Desensitising Agent. *Adv. Oral. Biol.* **2015**, *5*, 34–38.
21. Tschoppe, P.; Zandim, D.L.; Martus, P.; Kielbassa, A.M. Enamel and Dentine Remineralization by Nano-Hydroxyapatite Toothpastes. *J. Dent.* **2011**, *39*, 430–437. [CrossRef]
22. Adebayo, O.; Burrow, M.; Tyas, M. An SEM evaluation of conditioned and bonded enamel following carbamide peroxide bleaching and casein phosphopeptide-amorphous calcium phosphate (CPP-ACP) treatment. *J. Dent.* **2009**, *37*, 297–306. [CrossRef] [PubMed]
23. Caruana, P.C.; Al Mulaify, S.; Moazzez, R.; Bartlett, D. The effect of casein and calcium containing paste on plaque pH following a subsequent carbohydrate challenge. *J. Dent.* **2009**, *37*, 522–526. [CrossRef]
24. Souza, R.O.A.; Lombardo, G.H.L.; Pereira, S.M.B.; Zamboni, S.C.; Valera, M.C.; Araujo, M.A.M.; Ozcan, M. Analysis of tooth enamel after excessive bleaching: A study using scanning electron microscopy and energy dispersive X-ray spectroscopy. *Int. J. Prosthodont.* **2010**, *23*, 29–32.
25. Chuenarrom, C.; Benjakul, P.; Daosodsai, P. Effect of indentation load and time on knoop and vickers microhardness tests for enamel and dentin. *Mater. Res.* **2009**, *12*, 473–476. [CrossRef]
26. Farooq, I.; Majeed, A.; Alshwaimi, E.; Almas, K. Efficacy of a Novel Fluoride Containing Bioactive Glass Based Dentifrice in Remineralizing Artificially Induced Demineralization in Human Enamel. *Fluoride* **2019**, *52*, 447–455.
27. George, L.; Baby, A.; Dhanapal, T.P.; Charlie, K.M.; Joseph, A.; Varghese, A.A. Evaluation and comparison of the microhardness of enamel after bleaching with fluoride free and fluoride containing carbamide peroxide bleaching agents and post bleaching anticay application: An in vitro study. *Contemp. Clin. Dent.* **2015**, *6*, S163–S166. [CrossRef]
28. Molaasadolah, F.; Eskandarion, S.; Ehsani, A.; Sanginan, M. In Vitro Evaluation of Enamel Microhardness after Application of Two Types of Fluoride Varnish. *J. Clin. Diagn. Res.* **2017**, *11*, ZC64–ZC66. [CrossRef] [PubMed]
29. Sorozini, M.; Perez, C.R.; Rocha, G.M. Enamel sample preparation for AFM: Influence on roughness and morphology. *Microsc. Res. Tech.* **2018**, *81*, 1071–1076. [CrossRef] [PubMed]
30. Damato, F.A.; Strang, R.; Stephen, K.W. Effect of Fluoride Concentration on Remineralization of Carious Enamel: An In Vitro Ph-Cycling Study. *Caries Res.* **1990**, *24*, 174–180. [CrossRef] [PubMed]
31. Marsh, P.D. Microbiologic aspects of dental plaque and dental caries. *Dent. Clin. N. Am.* **1999**, *43*, 599–614.
32. Ali, S.; Farooq, I.; Al-Thobity, A.M.; Al-Khalifa, K.S.; Alhooshani, K.; Sauro, S. An in-vitro evaluation of fluoride content and enamel remineralization potential of two toothpastes containing different bioactive glasses. *Bio-Med. Mater. Eng.* **2020**, *30*, 487–496. [CrossRef]
33. Balakrishnan, A.; Jonathan, R.; Benin, P.; Kuumar, A. Evaluation to determine the caries remineralization potential of three dentifrices: An in vitro study. *J. Conserv. Dent.* **2013**, *16*, 375–379. [CrossRef]
34. Soares, R.; Ataide, I.D.N.D.; Fernandes, M.; Lambor, R. Assessment of Enamel Remineralisation after Treatment with Four Different Remineralising Agents: A Scanning Electron Microscopy (SEM) Study. *J. Clin. Diagn. Res.* **2017**, *11*, ZC136–ZC141. [CrossRef] [PubMed]

35. Hegde, M.N.; Moany, A. Remineralization of Enamel Subsurface Lesions with Casein Phosphopeptide-Amorphous Calcium Phosphate: A Quantitative Energy Dispersive X-Ray Analysis Using Scanning Electron Microscopy: An In Vitro Study. *J. Conserv. Dent.* **2012**, *15*, 61–67. [CrossRef] [PubMed]
36. Wegehaupt, F.J.; Tauböck, T.T.; Stillhard, A.; Schmidlin, P.R.; Attin, T. Influence of extra- and intra-oral application of CPP-ACP and fluoride on re-hardening of eroded enamel. *Acta Odontol. Scand.* **2011**, *70*, 177–183. [CrossRef] [PubMed]
37. Vyavhare, S.; Sharma, D.S.; Kulkarni, V.K. Effect of Three Different Pastes on Remineralization of Initial Enamel Lesion: An in Vitro Study. *J. Clin. Pediatr. Dent.* **2015**, *39*, 149–160. [CrossRef] [PubMed]
38. Cai, F.; Shen, P.; Morgan, M.V.; Reynolds, E.C. Remineralization of Enamel Subsurface Lesions In Situ by Sugar-Free Lozenges Containing Casein Phosphopeptide-Amorphous Calcium Phosphate. *Aust. Dent. J.* **2003**, *48*, 240–243. [CrossRef] [PubMed]
39. Jayarajan, J.; Janardhanam, P.; Jayakumar, P. Deepika Efficacy of CPP-ACP and CPP-ACPF on enamel remineralization—An in vitro study using scanning electron microscope and DIAGNOdent®. *Indian J. Dent. Res.* **2011**, *22*, 77–82. [CrossRef]
40. Somani, R.; Jaidka, S.; Singh, D.J.; Arora, V. Remineralizing potential of various agents on dental erosion. *J. Oral Biol. Craniofacial Res.* **2014**, *4*, 104–108. [CrossRef]
41. Juntavee, A.; Juntavee, N.; Hirunmoon, P. Remineralization Potential of Nanohydroxyapatite Toothpaste Compared with Tricalcium Phosphate and Fluoride Toothpaste on Artificial Carious Lesions. *Int. J. Dent.* **2021**, *2021*, 5588832. [CrossRef]
42. Amaechi, B.T.; AbdulAzees, P.A.; Alshareif, D.O.; Shehata, M.A.; Lima, P.P.D.C.S.; Abdollahi, A.; Kalkhorani, P.S.; Evans, V. Comparative efficacy of a hydroxyapatite and a fluoride toothpaste for prevention and remineralization of dental caries in children. *BDJ Open* **2019**, *5*, 1–9. [CrossRef]
43. Najibfard, K.; Ramalingam, K.; Chedjieu, I.; Amaechi, B.T. Remineralization of early caries by a nano-hydroxyapatite dentifrice. *J. Clin. Dent.* **2011**, *22*, 139–143. [PubMed]
44. Grandjean, P. Developmental fluoride neurotoxicity: An updated review. *Environ. Health* **2019**, *18*, 1–17. [CrossRef]
45. Huang, S.B.; Gao, S.S.; Yu, H.Y. Effect of nano-hydroxyapatite concentration on remineralization of initial enamel lesion in vitro. *Biomed. Mater.* **2009**, *4*, 034104. [CrossRef] [PubMed]
46. Grocholewicz, K.; Matkowska-Cichocka, G.; Makowiecki, P.; Droździk, A.; Ey-Chmielewska, H.; Dziewulska, A.; Tomasik, M.; Trybek, G.; Janiszewska-Olszowska, J. Effect of nano-hydroxyapatite and ozone on approximal initial caries: A randomized clinical trial. *Sci. Rep.* **2020**, *10*, 1–8. [CrossRef] [PubMed]
47. Shaik, Z.A.; Rambabu, T.; Sajjan, G.; Varma, M.; Satish, K.; Raju, V.B.; Ganguru, S.; Ventrapati, N. Quantitative Analysis of Remineralization of Artificial Carious Lesions with Commercially Available Newer Remineralizing Agents Using SEM-EDX- In Vitro Study. *J. Clin. Diagn. Res.* **2017**, *11*, ZC20–ZC23. [CrossRef]
48. Wang, Z.; Jiang, T.; Sauro, S.; Pashley, D.H.; Toledano, M.; Osorio, R.; Liang, S.; Xing, W.; Sa, Y.; Wang, Y. The dentine remineralization activity of a desensitizing bioactive glass-containing toothpaste: An in vitro study. *Aust. Dent. J.* **2011**, *56*, 372–381. [CrossRef]

Article

Treatment of Intrabony Defects with a Combination of Hyaluronic Acid and Deproteinized Porcine Bone Mineral

Darko Božić [1,*], Ivan Ćatović [2], Ana Badovinac [1], Larisa Musić [1], Matej Par [3] and Anton Sculean [4]

1. Department of Periodontology, School of Dental Medicine, University of Zagreb, HR-10000 Zagreb, Croatia; badovinac@sfzg.hr (A.B.); lmusic@sfzg.hr (L.M.)
2. Private Dental Practice, HR-52100 Pula, Croatia; ivan.catovic@gmail.com
3. Department of Endodontics and Restorative Dentistry, School of Dental Medicine, University of Zagreb, HR-10000 Zagreb, Croatia; mpar@sfzg.hr
4. Department of Periodontology, School of Dental Medicine, University of Bern, CH-3010 Bern, Switzerland; anton.sculean@zmk.unibe.ch
* Correspondence: bozic@sfzg.hr; Tel.: +385-1480-2155

Abstract: Background: this study evaluates the clinical outcomes of a novel approach in treating deep intrabony defects utilizing papilla preservation techniques with a combination of hyaluronic acid (HA) and deproteinized porcine bone mineral. Methods: 23 patients with 27 intrabony defects were treated with a combination of HA and deproteinized porcine bone mineral. Clinical attachment level (CAL), pocket probing depth (PPD), gingival recession (REC) were recorded at baseline and 6 months after the surgery. Results: At 6 months, there was a significant CAL gain of 3.65 ± 1.67 mm ($p < 0.001$) with a PPD reduction of 4.54 ± 1.65 mm ($p < 0.001$), which was associated with an increase in gingival recession (0.89 ± 0.59 mm, $p < 0.001$). The percentage of pocket resolution based on a PPD ≤ 4 mm was 92.6% and the failure rate based on a PPD of 5 mm was 7.4%. Conclusions: the present findings indicate that applying a combined HA and xenograft approach in deep intrabony defects provides clinically relevant CAL gains and PPD reductions compared to baseline values and is a valid new approach in treating intrabony defects.

Keywords: hyaluronic acid; periodontal regeneration; intrabony defects; microsurgery; periodontitis; surgical flaps

1. Introduction

Periodontitis is a non-communicable chronic inflammatory disease caused by periodontal pathogenic biofilm [1–4] and with 743 million affected people worldwide in its severe form, it is the sixth most prevalent disease globally [5,6]. If left untreated, the disease leads to tooth loss and has a significant social and economic impact [7].

Following non-surgical therapy, it is common that some deep intrabony defects remain, presenting an increased risk of disease progression with further attachment loss that could lead to tooth loss [8,9].

Therefore, one of the ultimate goals in periodontology is to achieve periodontal regeneration and change the prognosis of questionable or hopeless teeth to maintainable. Over the last several decades, periodontal regenerative procedures have seen a change in flap designs and materials used to promote periodontal regeneration [10]. Human histological studies have shown that different materials can promote periodontal regeneration to various success [11]. In addition, several clinical studies have shown long-term stability of clinical attachment gain of periodontally compromised teeth when treated with regenerative procedures benefitting the patient in retaining their teeth [12–16].

Results from systematic reviews evaluating the clinical outcomes obtained with various biologics/growth factors for regenerative periodontal therapy have shown positive clinical effects evidenced by the gain of clinical attachment and pocket-depth reduction [17,18].

Citation: Božić, D.; Ćatović, I.; Badovinac, A.; Musić, L.; Par, M.; Sculean, A. Treatment of Intrabony Defects with a Combination of Hyaluronic Acid and Deproteinized Porcine Bone Mineral. *Materials* **2021**, *14*, 6795. https://doi.org/10.3390/ma14226795

Academic Editor: Eugenio Velasco-Ortega

Received: 6 October 2021
Accepted: 5 November 2021
Published: 11 November 2021

Publisher's Note: MDPI stays neutral with regard to jurisdictional claims in published maps and institutional affiliations.

Copyright: © 2021 by the authors. Licensee MDPI, Basel, Switzerland. This article is an open access article distributed under the terms and conditions of the Creative Commons Attribution (CC BY) license (https://creativecommons.org/licenses/by/4.0/).

Based on the accumulating evidence on the benefits of periodontal regenerative procedures and the current evidence related to the use of these procedures, very recently Nibali pointed out that periodontal regenerative procedures should be the treatment of choice for intrabony defects [19].

Over the last several years, another emerging molecule serving as a potential candidate for periodontal regeneration is the hyaluronic acid (HA), a key extracellular matrix component involved in cell migration. As a major component of the extracellular matrix, it is expressed in various cells of the periodontium [20]. Furthermore, its main receptor, CD44, a cell surface molecule, is expressed by PDL cells and cementoblasts [21–24]. Of importance for periodontal regeneration, it has been shown that the interaction between CD44-HA in PDL cells is critical for the proliferation and migration of these cells [23,24]. Furthermore, data from other in vitro studies revealed that HA induces early osteogenic differentiation of PDL cells [25], increases the migratory and proliferative properties of gingival fibroblasts [26] and maintains the stemness of mesenchymal stromal cells and pre-osteoblasts, making it a valid candidate for bone and periodontal regeneration [27]. In addition, HA can induce proliferation and osteogenic differentiation of bone marrow stromal cells and pre-osteoblastic cells, having an important role in the early and late stages of bone formation [28,29]. Recent animal studies have shown an increase in bone formation and improved wound healing after applying HA in chronic pathology-type extraction sockets, comparable to that observed following the use of BMP-2 [30,31]. These findings are in line with those from two animal studies evaluating the effects of HA in the treatment of intrabony and recession defects revealing periodontal regeneration evidenced through the formation of cementum, periodontal ligament and bone [32,33].

Thus, taken together, the current evidence suggests that HA may not only positively influence periodontal regeneration but may also have a potential role in bone formation. Indeed, recent systematic reviews have shown that the adjunctive use of HA in both non-surgical and surgical periodontal therapy may have a positive effect on the clinical outcomes evidenced by pocket probing depth (PPD) reduction, clinical attachment level (CAL) gain and reduction of bleeding on probing (BOP). However, the authors also suggested a need for further studies that would investigate HA in different clinical scenarios [34,35].

A recent randomized clinical study with a 24-month follow-up provided further evidence of a positive effect HA has on CAL gains when enamel matrix derivative (EMD) was compared with HA in the surgical treatment of intrabony defects. The study showed that both materials may lead to comparable clinical results in terms of PPD reductions and CAL gains [36]. To further improve the clinical outcomes of regenerative therapy by stabilizing the blood clot and providing space for periodontal regeneration, bone replacement grafts were combined with an enamel matrix derivative (EMD). The combination of EMD and bone replacement grafts resulted in an additional gain of CAL (i.e., 1mm) when compared to the use of EMD alone [37]. Interestingly, when HA was used alone or combined with a volume stable collagen matrix, the combination approach yielded a higher but not statistically significant amount of new cementum, new periodontal ligament and new bone compared to the use of HA alone [33]. These very recent findings suggest that a filler material might be beneficial for periodontal regeneration when using HA, thus stabilizing the blood clot and providing flap support to prevent its collapse into the intrabony defect and creating more space for periodontal regeneration.

However, no clinical data are currently available evaluating the combined use of HA with a bone replacement graft in intrabony defects. Therefore, this pilot case series aimed to clinically evaluate, for the first time, the healing of intrabony defects treated with a combination of HA and a bone replacement material.

2. Materials and Methods

2.1. Study Setting and Ethical Considerations

The study was conducted at the Department of Periodontology, School of Dental Medicine, University of Zagreb, between June 2019 and December 2020. All clinical procedures were performed in full accordance with the Declaration of Helsinki and the Good Clinical Practice Guidelines. Each patient provided written informed consent. The study protocol was approved by the Ethical committee of the School of Dental Medicine University of Zagreb, Croatia (05-PA-30-XXI).

2.2. Study Population

Patients referred to the Department of Periodontology for periodontal treatment were consecutively screened for study inclusion. Before enrolling, all patients underwent cause-related therapy, consisting of oral hygiene instructions and scaling and root planing with machine-driven and hand instruments. Splinting of mobile teeth was performed if necessary.

The inclusion criteria were as follows: (i) diagnosis of stage III or IV periodontitis; (ii) good general health with no systemic diseases that could contraindicate surgery, no medications that could affect the periodontal status, uncontrolled or poorly controlled diabetes, no pregnancy or lactation; (iii) patients had to have at least one intrabony defect with PPD $\geq$ 6 mm, CAL $\geq$ 6 mm and an intrabony component $\geq$ 4 mm measured on digital periapical radiographs that predominantly involved the interproximal area of the affected tooth; (iv) FMPS and FMBS $\leq$ 20% following non-surgical treatment [38]; (v) vital teeth or teeth with properly performed endodontic treatment.

Exclusion criteria were: (i) teeth with degree III mobility, furcation involvement, or inadequate endodontic treatment and/or restoration; (ii) heavy smokers (more than 10 cig/day). We enrolled 23 systemically healthy patients in this case-series study (16 females and 7 males, mean age 54.59 $\pm$ 10.24, age range: 35–85 years), 4 of whom were smokers (<10 cigarettes per day).

2.3. Surgical Procedures

All surgical procedures were performed by an experienced periodontist (D.B.) with more than 20 years of clinical experience. Following local anesthesia, the defect-associated papillary area was either accessed with the simplified papilla preservation flap (SPPF) [39] or with the modified papilla preservation flap (MPPT) [40]. The flap design was chosen based on the width of the interdental space; when 2 mm or less SPPF was utilized, while when the interdental space was wider than 2 mm MPPT was used.

The incisions were intrasulcular in order to preserve the width and the height of the defect-associated papilla. The mesial-distal extension was at least one tooth mesial and distal to the defect site in order to provide access to the base of the bony defect and allow proper visualization and debridement of the defect. Periosteal incisions were never performed.

Following a full-thickness flap reflection, granulation tissue was removed from the intrabony defect using curettes and microscissors. Scaling and root planing of the root surface was performed with both hand curettes (Gracey, Hu-Friedy, Chicago, IL, USA) and power-driven instruments (SONICflex LUX, KaVo Dental GmbH, Biberach, Germany).

2.4. Application of the Hyaluronic Acid (HA) and Xenograft

At the end of the instrumentation, the defect was rinsed with sterile saline and EDTA (sterile 24% EDTA gel, pH 6.7; PrefGel, Straumann, Basel, Switzerland) was applied on the instrumented root surface for 2 min. The defect area was then carefully rinsed with saline and a thin layer of cross-linked HA gel (HYADENT BG®, BioScience, Germany: a gel formulation containing cross-linked HA (1000 kDA HA monomers) and non-cross-linked HA (2500 kDA) in a ratio 8:1, made from biotechnologically produced synthetic HA) was applied on the root surface. Afterwards a 1:1 ratio mixture of HA and deproteinized porcine bone mineral (THE Graft, Purgo Biologics Inc., Korea) was gently packed into the intrabony defect to the level of the bone, not overfilling the defect. The flaps were repositioned and primary wound closure was achieved with a horizontal internal mattress suture at the base of the papilla and a single interrupted suture to connect the tips of the papillae. The papillae were sutured using the monofilament non-resorbable 5-0/6-0 suturing material (Ethilon, Ethicon, Johnson and Johnson, Somerville, NJ, USA) (Figures 1 and 2).

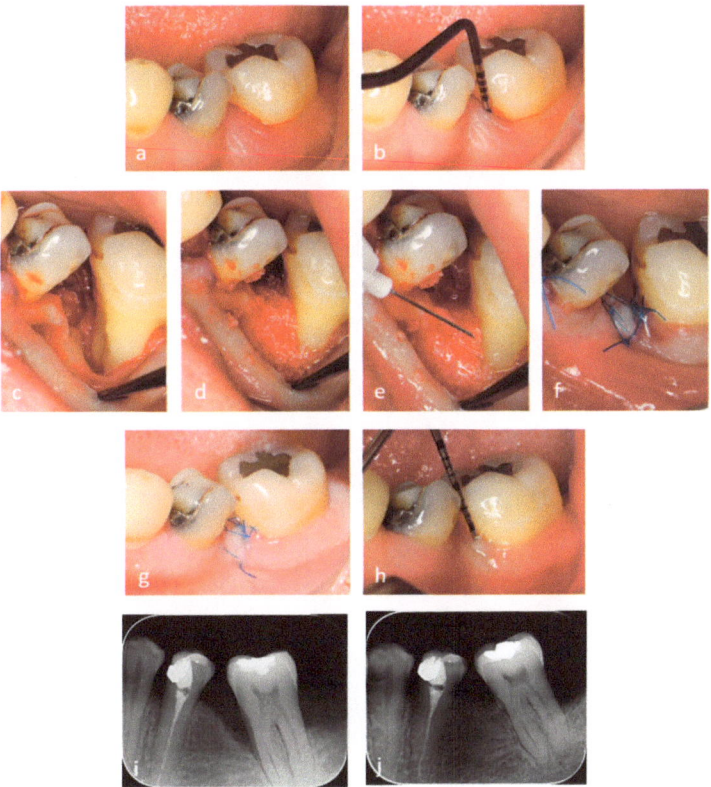

Figure 1. (**a**,**b**) Preoperative view and a pocket probing depth (PPD) of 9 mm. (**c**) Clinical view of the 1-wall intrabony defect and the instrumented root surface with a thin layer of hyaluronic acid (HA) applied. (**d**) Mixture of HA and xenograft gently packed in the intrabony defect. (**e**) HA layer applied before suturing. (**f**) Passive primary wound closure. (**g**) 14 days healing before suture removal. (**h**) PPD of 3 mm at 6 months. (**i**) Pre-operative radiographic finding. (**j**) Radiographic finding at 6 months.

2.5. Post-Surgical Instructions and Plaque Control

Patients received systemic antibiotic therapy amoxicillin + clavulanic acid 1 g/day for seven days. Pain control was obtained by 400 mg ibuprofen three times per day for the first 24 h and subsequently based on the patient's need. Each patient was advised to rinse twice per day with a 0.2% chlorhexidine gluconate solution (Parodontax extra, GSK, Brentford, UK) for 4 weeks. Smoking patients were asked to refrain from smoking during the first 4 post-operative weeks. Sutures were removed 14 days following surgery, and the patients were instructed to brush with a post-surgical soft toothbrush. The use of a soft toothbrush was discontinued after 3 months when a medium bristled toothbrush was re-introduced. Each patient received professional tooth cleaning during the monthly control appointments for the following 6 months.

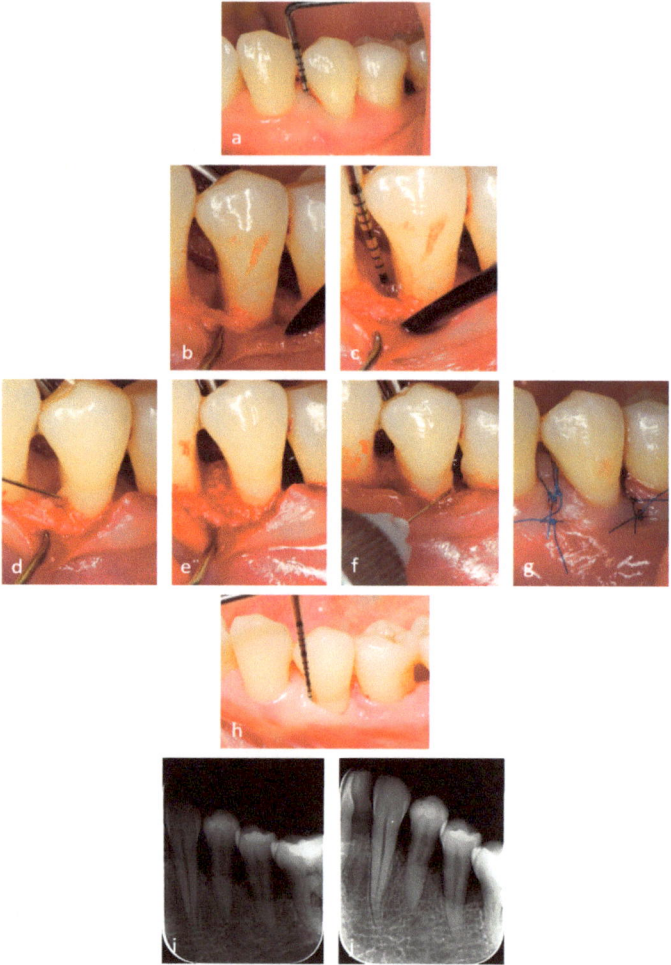

Figure 2. (**a**) Preoperative view and PPD depth of 8 mm. (**b**,**c**) Clinical intraoperative view of the defect with a 7mm depth of the intrabony component. (**d**) Clinical view of the instrumented root surface with a thin layer of HA applied. (**e**) Mixture of HA and xenograft gently packed in the intrabony defect. (**f**) HA layer applied before suturing on the surgical area. (**g**) Passive primary wound closure. (**h**) PPD of 2 mm at 6 months. (**i**) Pre-operative radiographic finding. (**j**) Radiographic finding at 6 months.

2.6. Outcome Measures

The primary outcome measure defined in the study was the change in CAL between baseline and 6 months. The secondary outcomes were changes in PPD and REC.

All clinical measurements were carried out by a single examiner (I.Ć.) at baseline and 6 months after surgery. Prior to the study, the examiner was calibrated to reduce the intraexaminer error (=0.893) in order to reach reliability and consistency. Pocket probing depths and gingival recessions were rounded to the nearest 0.5 mm at the deepest part of the interproximal site. Clinical attachment level was calculated as a sum of PPD and REC. For the first month after the surgery, the primary closure of the surgical sites was evaluated on a weekly basis.

2.7. Statistical Analysis

The assumption of normality of distribution was evaluated using Shapiro-Wilk's test and the inspection of normal Q-Q plots. Homogeneity of variances was verified using Levene's test. For PPD, CAL, and REC, baseline values were compared to the values measured after 6 months using paired t-tests. The relationships for binary combinations of outcome variables were explored using Pearson's correlation analysis, except for the variable EHI, which was analyzed using Spearman's correlation due to the violation of normality. p-values lower than 0.05 were considered statistically significant. The statistical analysis was performed using SPSS (version 25, IBM, Armonk, NY, USA).

3. Results

3.1. Baseline Data

A total of 27 defects were treated with 17 intrabony defects located in the maxilla, and 10 were in the mandible (8 incisors, 3 canines, 9 premolars, 7 molars).

The mean distance from the cementoenamel junction (CEJ) to the bottom of the defect (CEJ-BD) was 10.54 ± 3.23 mm, and the mean intrabony component (IB) was 7.24 ± 2.46 mm. The mean intrabony width of the osseous defect (IBW) was 3.44 ± 0.96, and the mean value of the radiographic defect angle was 26.4 ± 8.43°. From 27 intrabony defects, 13 were 2-wall, 7 were 1-wall, 6 were 3-wall and one was a crater defect. Patient and defect characteristics are presented in Table 1.

Table 1. Patient and defect characteristics (respectively, n = 23 and n = 27).

Study population	
Age (years, mean ± SD)	54.59 ± 10.24
Gender (male/female)	7/16
Smoking (yes/no)	4/19
Defect characteristics	
Dental arch (maxillary/mandibular)	17/10
Tooth type (incisor/canine/premolar/molar)	8/3/9/7
CEJ-defect bottom (mean ± SD, mm)	10.54 ± 3.23
Intrabony component (mean ± SD, mm)	7.24 ± 2.46
Intrabony width (mean ± SD, mm)	3.44 ± 0.96
X-ray angle (mean ± SD, degree)	26.4 ± 8.43
Defect configuration	
1-wall	7
2-wall	13
3-wall	6
Crater	1

CEJ, cementoenamel junction; SD, standard deviation.

3.2. Early Wound Healing

Primary closure of the incision lines was achieved in 100% of the cases. We assessed the early wound healing index (EHI) during wound healing [41]. At the 2-week suture removal time of the 27 treated sites, 13 sites had an EHI score of 1, 10 sites had EHI 2, 3 sites had an EHI score of 4, and 1 site had a score of 3. Spearman's correlation analysis found no significant relationship between EHI score and CAL gains and PPD reductions.

3.3. Clinical Outcomes at 6 Months

Six months after the surgery, the results showed statistically significant changes for PPD, CAL and REC ($p < 0.001$). The mean residual PPD after 6 months was 3.35 ± 0.72 mm with a decrease of 4.54 ± 1.65 mm. Of the 27 treated sites 14 sites had a residual PPD of 2–3 mm, 11 sites had 4 mm and 2 sites had a PPD of 5 mm. The CAL also significantly changed from baseline to 6 months with an average CAL gain of 3.65 ± 1.67 mm. Twelve sites showed CAL gains of 2–3 mm, and 3 sites had 4mm of CAL gain and 11 sites reached CAL gains of ≥ 5 mm. At 6 months, there was a mean increase of 0.89 ± 0.59 mm in REC compared with the baseline value (0.83 ± 0.67 mm). Clinical baseline data and outcomes 6 months after treatment, and the frequency distribution of CAL gains, residual PPD and REC after 6 months are shown in Tables 2 and 3.

Table 2. Clinical outcomes at baseline and 6 months after treatment ($n = 27$).

Variable	Baseline	6 Months	Change	Significance, p [a]
PPD (mm)	7.89 ± 1.60	3.35 ± 0.72	4.54 ± 1.65	<0.001
CAL (mm)	8.72 ± 1.82	5.07 ± 1.28	3.65 ± 1.67	<0.001
REC (mm)	0.83 ± 0.67	1.72 ± 0.90	0.89 ± 0.59	<0.001

[a] Paired t-test. PPD, pocket probing depth; CAL, clinical attachment level; REC, gingival recession.

Table 3. Frequency distribution of CAL gains and residual PPD after 6 months.

	CAL Gain		Residual PPD	
	n	%	n	%
0–1 mm	1	3.7	0	0
2–3 mm	12	44.4	14	51.9
4 mm	3	11.1	11	40.7
5 mm	6	22.2	2	7.4
≥ 6 mm	5	18.5	0	0

CAL, clinical attachment level; PPD, pocket probing depth.

A separately conducted statistical analysis was done for the four smoking patients for CAL gains, PPD reductions and REC changes and no statistical significance was found compared to the non-smoking patients.

Correlations among interval variables were analyzed and CAL gain significantly and positively correlated with preoperative PPD and CAL (respectively: $r = 0.812$, $p < 0.001$, $r = 0.731$, $p < 0.001$), PPD reduction ($r = 0.936$, $p < 0.001$), IB component ($r = 0.494$, $p < 0.009$) and CEJ-BD ($r = 0.526$, $p < 0.005$).

Similar correlations were found for PPD reduction, where PPD reduction strongly correlated with preoperative PPD and CAL (respectively: $r = 0.903$, $p < 0.001$, $r = 0.815$, $p < 0.001$), CAL gain ($r = 0.936$, $p < 0.001$), IB component ($r = 0.484$, $p < 0.01$) and CEJ-BD ($r = 0.601$, $p < 0.001$).

The correlations with the highest Pearson's coefficient are shown in Figure 3.

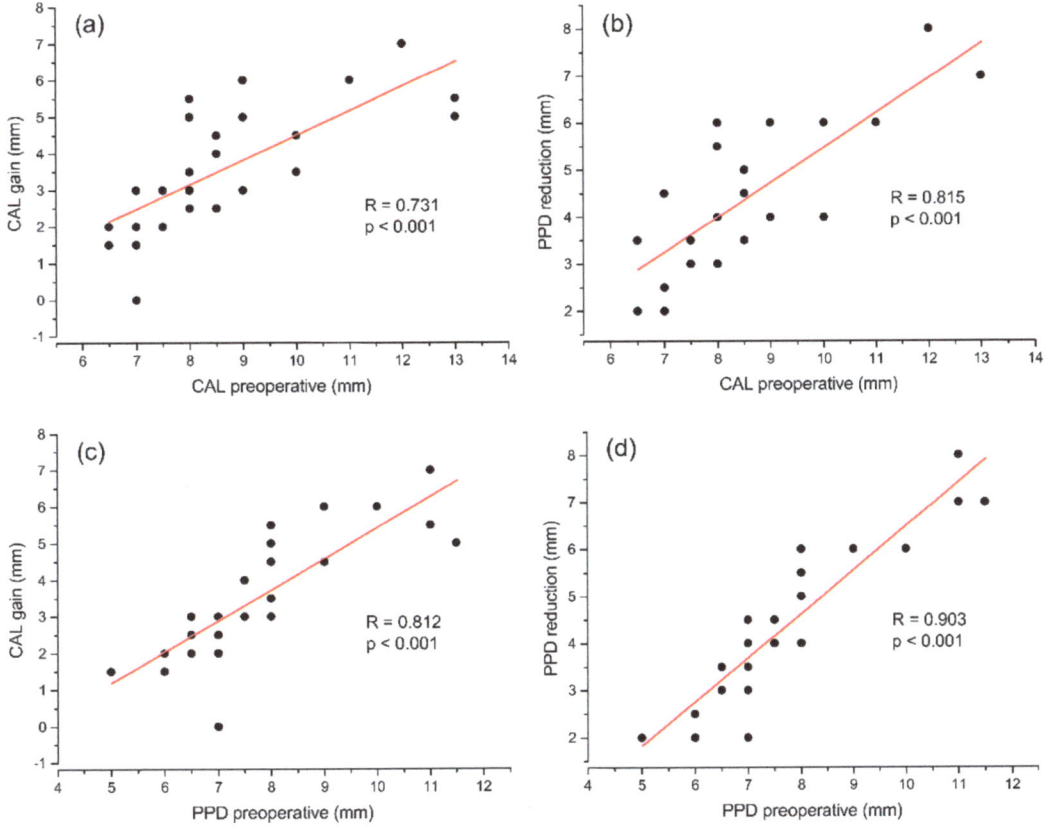

Figure 3. Correlations among interval variables. (**a**) CAL gain compared to preoperative CAL. (**b**) PPD reduction compared to preoperative CAL. (**c**) CAL gain compared to preoperative PPD. (**d**) PPD reduction compared to preoperative PPD.

4. Discussion

This case-series study shows that the combination of HA and a xenograft can yield substantial clinical improvements in deep interdental intrabony defects. Utilizing well-established surgical techniques to obtain primary closure and wound stabilization, the results of this study appear to point to the potential relevance of HA to achieve clinically relevant improvements in terms of CAL gains and PPD reductions in deep intrabony defects.

HA is a molecule that has been reported to stimulate the proliferation of gingival fibroblasts, PDL cells, induce osteogenic differentiation of bone marrow stromal cells (BMSC) and is well known for promoting angiogenesis and neovascularization [26,29,42]. Recent histological animal studies that evaluated HA for periodontal regeneration confirmed that the application of HA can promote periodontal regeneration in 2-wall and dehiscence defects [32,33], possibly by interacting with CD44 during the early phases of periodontal tissue regeneration [43].

In recent decades' various biological agents, either alone or combined with bone grafts, have gained much interest in regenerative periodontal surgery, and clinical trials have shown their efficacy in achieving significant CAL gains and PPD reductions [17,44–46]. All studies have shown that when biologics are combined with bone grafts greater CAL gains and PPD reductions are achieved, further improving clinical outcomes.

Recently HA has gained attraction with several clinical trials conducted to assess its efficacy in periodontal regeneration [36,47,48]. Of particular interest to our results is the study by Pilloni since they used the same hyaluronic acid as we did. In their randomized clinical trial comparing EMD to HA over 24 months, they achieved CAL gains of 2.19 ± 1.11 mm in the HA group and 2.94 ± 1.12 mm in the EMD group and PD reductions of 4.5 ± 0.97 mm for EMD and 3.31 ± 0.70 mm for HA.

Comparing the present results to those mentioned above, it appears that the use of HA combined with a bone grafting material may further improve the clinical outcomes compared to the use of HA alone (i.e., CAL gains of 3.65 ± 1.67 mm vs. 2.19 ± 1.11 mm, and PD reductions of 4.54 ± 1.65 mm vs. 3.31 ± 0.70 mm). One of the reasons for these observed differences could be due to the fact that HA, similar to other biologics, has a fluid consistency which then prevents it to possess a sufficient space-making potential needed for periodontal regeneration. This could then lead to the collapse of the mucoperiosteal flap and subsequently limit the outcomes of regenerative surgery. By combining a bone-grafting material with HA this could then provide sufficient support to the mucoperiosteal flap preventing its collapse into the defect, thus stabilizing the blood clot and providing additional space needed for periodontal regeneration. These results are in line with those of a recent systematic review that analyzed the application of EMD with and without bone grafts (46). When EMD was combined with bone grafts the CAL gains were 3.76 ± 1.07 mm compared to 3.32 ± 1.04 mm when EMD was used alone while the PPD reductions were 4.22 ± 1.20 mm in the combination group compared to 4.12 ± 1.07 mm when EMD was alone. However, the increase in REC amounted 0.76 ± 0.42 mm, which is comparable to our value of 0.89 ± 0.59 mm. The reason for such an increase in the recession in our study could be due to the fact that the majority of our defects were 1- and 2-wall or combinations of the two with little support of the surrounding tissues. This assumption is in line with those of a previous study by De Leonardis and Paolantonio where they reported CAL gains of 3.47 ± 0.65mm and PPD reductions of 4.00 ± 0.42 mm following the combination of EMD with a bone graft in 1- and 2-wall defects [49]. The importance of intrabony defect morphology, number of walls, radiographic angle and depth of the intrabony component was recently analyzed in a systematic review and meta-analysis [50]. The authors concluded that these parameters affect the clinical outcomes, indicating that deeper, narrower defects and defects with more walls are associated with improved clinical and radiographic outcomes. To overcome the issue of a negative effect of a defect morphology on the clinical outcome, the additional use of filler material is recommended in these non-space maintaining defects to increase the stability of the blood clot and create space for the regeneration process. In this context, the present study results in terms of CAL gains and PPD reductions are comparable to those of other studies that have utilized different growth factors combined with bone grafting materials, thus supporting the use of combination approaches for periodontal regeneration [17,44–46].

However, CAL gain changes alone have to be considered carefully, especially if periodontal pockets deeper than 4mm still remain since they may represent a significant risk factor for long-term disease progression/recurrence [9]. Therefore, in a recent publication, Trombelli et al. [51] proposed a composite outcome of combined CAL gains and PPD ≤ 4 mm examining the frequency of these clinically relevant outcomes. Furthermore, in a recent systematic review by Aimetti et al. the authors assessed the frequency of reported pocket resolution (PPD ≤ 3 mm and PPD ≤ 4 mm) outcomes in the regenerative treatment of intrabony defects utilizing papilla preservations techniques (PPTs) [52]. Their final analysis included 12 randomized clinical trials and found that the pocket resolution with a PPD ≤ 3 mm was achieved in 61.4% and with a PPD ≤ 4 mm in 92.1%. However, for the PPD ≤ 3 mm, the results ranged from 28.6% to 93.3%, while for PPD ≤ 4 mm, the range was less variable, ranging from 71.4% to 100%. In the present study, we found that in 51.9% of the sites we had a remaining PPD ≤ 3 mm, and in another 40.7% of the sites a PPD of 4 mm, with two sites having a PPD of 5 mm, 7.4%. However, when we consider PPD of 4mm as the cut-off value, the overall percentage of pocket resolution is 92.6%, which is in line with the aforementioned systematic review. When we compare the 6 months results from Trombelli et al. [51] to our 6 months' results, similarities in terms of CAL gains, pocket closure percentages and failure rates, defined as PPD > 4 mm, are observed. CAL gains in the Trombelli study were 3.7 mm, compared to our 3.65 mm, the average residual PPDs were 3.7 mm compared to our 3.35 mm, and failure rates indicated by a PPD > 4 mm were 7% compared to our 7.4%. When we look at pocket closure percentages, the authors reported 79.6% compared to 92.6% in our study. This discrepancy could be due to the bigger number of treated sites in their study n = 103, compared to n = 27 in our study. However, these results show that the combined approach using HA + xenograft with PPTs is able to achieve clinically relevant PPD reductions, CAL gains and pocket resolution compared to other clinical studies.

Limitations of the present study are the lack of a control group treated with HA alone and the short-term follow-up of 6 months, although there are a number of multicenter clinical trials with 6 months of follow-up [44,45,51]. Furthermore, although radiographs were taken, they were not standardized and, therefore, no precise hard tissue measurements were possible.

5. Conclusions

In conclusion, the application of a combined approach of HA and a xenograft in deep intrabony defects provides clinically relevant CAL gains and PPD reductions with over 90% of pocket closure based on a PPD of 4 mm. It can thus be anticipated that this approach may represent a new treatment option for deep intrabony defects. Therefore, randomized clinical trials comparing the use of HA alone with HA plus bone graft are warranted to shed light on the clinical relevance of this novel treatment approach.

Author Contributions: Conceptualization, D.B.; methodology, D.B.; data collection, interpretation I.Ć., A.B., L.M., M.P. and D.B.; formal analysis, M.P.; investigation, D.B. and I.Ć.; resources, D.B.; data curation, I.Ć. and M.P.; writing—original draft preparation, D.B. and A.B.; writing—review and editing, D.B. and A.S.; visualization, D.B., A.B., L.M. and M.P.; supervision, D.B. and A.S.; project administration, D.B.; funding acquisition, D.B. All authors have read and agreed to the published version of the manuscript.

Funding: This research received no external funding. However, it was partially supported with materials, the Hyaluronic acid and THE Graft were kindly provided free of charge by Regedent (Zurich, Switzerland).

Institutional Review Board Statement: The study was conducted according to the guidelines of the Declaration of Helsinki, and approved by the Ethics Committee of the School od Dental Medicine University of Zagreb (05-PA-30-XXI).

Informed Consent Statement: Informed consent was obtained from all subjects involved in the study.

Data Availability Statement: The data presented in this study are available on request from the corresponding author.

Acknowledgments: The authors would like to thank Regedent for the kind donation of materials.

Conflicts of Interest: The authors declare no conflict of interest.

References

1. Abusleme, L.; Hoare, A.; Hong, B.; Diaz, P.I. Microbial Signatures of Health, Gingivitis, and Periodontitis. *Periodontol. 2000* **2021**, *86*, 57–78. [CrossRef]
2. Darveau, R.P.; Curtis, M.A. Oral Biofilms Revisited: A Novel Host Tissue of Bacteriological Origin. *Periodontol. 2000* **2021**, *86*, 8–13. [CrossRef] [PubMed]
3. Jakubovics, N.S.; Goodman, S.D.; Mashburn-Warren, L.; Stafford, G.P.; Cieplik, F. The Dental Plaque Biofilm Matrix. *Periodontol. 2000* **2021**, *86*, 32–56. [CrossRef] [PubMed]
4. Joseph, S.; Curtis, M.A. Microbial Transitions from Health to Disease. *Periodontol. 2000* **2021**, *86*, 201–209. [CrossRef]
5. Marcenes, W.; Kassebaum, N.J.; Bernabé, E.; Flaxman, A.; Naghavi, M.; Lopez, A.; Murray, C.J.L. Global Burden of Oral Conditions in 1990-2010: A Systematic Analysis. *J. Dent. Res.* **2013**, *92*, 592–597. [CrossRef] [PubMed]
6. Kassebaum, N.J.; Bernabé, E.; Dahiya, M.; Bhandari, B.; Murray, C.J.L.; Marcenes, W. Global Burden of Severe Periodontitis in 1990-2010: A Systematic Review and Meta-Regression. *J. Dent. Res.* **2014**, *93*, 1045–1053. [CrossRef] [PubMed]

7. Jin, L.; Lamster, I.; Greenspan, J.; Pitts, N.; Scully, C.; Warnakulasuriya, S. Global Burden of Oral Diseases: Emerging Concepts, Management and Interplay with Systemic Health. *Oral Dis.* **2016**, *22*, 609–619. [CrossRef] [PubMed]
8. Papapanou, P.N.; Wennstrom, J.L. The Angular Bony Defect as Indicator of Further Alveolar Bone Loss. *J. Clin. Periodontol.* **1991**, *18*, 317–322. [CrossRef] [PubMed]
9. Matuliene, G.; Pjetursson, B.E.; Salvi, G.E.; Schmidlin, K.; Brägger, U.; Zwahlen, M.; Lang, N.P. Influence of Residual Pockets on Progression of Periodontitis and Tooth Loss: Results after 11 Years of Maintenance. *J. Clin. Periodontol.* **2008**, *35*, 685–695. [CrossRef]
10. Cortellini, P.; Tonetti, M.S. Clinical Concepts for Regenerative Therapy in Intrabony Defects. *Periodontol. 2000* **2015**, *68*, 282–307. [CrossRef]
11. Sculean, A.; Nikolidakis, D.; Nikou, G.; Ivanovic, A.; Chapple, I.L.C.; Stavropoulos, A. Biomaterials for Promoting Periodontal Regeneration in Human Intrabony Defects: A Systematic Review. *Periodontol. 2000* **2015**, *68*, 182–216. [CrossRef] [PubMed]
12. Cortellini, P.; Buti, J.; Pini Prato, G.; Tonetti, M.S. Periodontal Regeneration Compared with Access Flap Surgery in Human Intra-Bony Defects 20-Year Follow-up of a Randomized Clinical Trial: Tooth Retention, Periodontitis Recurrence and Costs. *J. Clin. Periodontol.* **2017**, *44*, 58–66. [CrossRef] [PubMed]
13. Silvestri, M.; Rasperini, G.; Milani, S. 120 Infrabony Defects Treated With Regenerative Therapy: Long-Term Results. *J. Periodontol.* **2011**, *82*, 668–675. [CrossRef] [PubMed]
14. De Ry, S.P.; Roccuzzo, A.; Lang, N.P.; Sculean, A.; Salvi, G.E. Long-Term Clinical Outcomes of Periodontal Regeneration with Enamel Matrix Derivative: A Retrospective Cohort Study with a Mean Follow-up of 10 Years. *J. Periodontol.* **2021**. [CrossRef] [PubMed]
15. Sculean, A.; Donos, N.; Schwarz, F.; Becker, J.; Brecx, M.; Arweiler, N.B. Five-Year Results Following Treatment of Intrabony Defects with Enamel Matrix Proteins and Guided Tissue Regeneration. *J. Clin. Periodontol.* **2004**, *31*, 545–549. [CrossRef]
16. Sculean, A.; Kiss, A.; Miliauskaite, A.; Schwarz, F.; Arweiler, N.B.; Hannig, M. Ten-Year Results Following Treatment of Intra-Bony Defects with Enamel Matrix Proteins and Guided Tissue Regeneration. *J. Clin. Periodontol.* **2008**, *35*, 817–824. [CrossRef]
17. Kao, R.T.; Nares, S.; Reynolds, M.A. Periodontal Regeneration—Intrabony Defects: A Systematic Review from the AAP Regeneration Workshop. *J. Periodontol.* **2015**, *86*, S77–S104. [CrossRef]
18. Li, F.; Yu, F.; Xu, X.; Li, C.; Huang, D.; Zhou, X.; Ye, L.; Zheng, L. Evaluation of Recombinant Human FGF-2 and PDGF-BB in Periodontal Regeneration: A Systematic Review and Meta-Analysis. *Sci. Rep.* **2017**, *7*, 65. [CrossRef]
19. Nibali, L. Guest Editorial: Time to Reflect on New Evidence about Periodontal Regenerative Surgery of Intrabony Defects. *J. Clin. Periodontol.* **2021**, *48*, 557–559. [CrossRef]
20. Bartold, P. Proteoglycans of the Periodontium: Structure, Role and Function—PubMed. *J. Periodontal Res.* **1987**, *22*, 431–444. [CrossRef]
21. Yeh, Y.; Yang, Y.; Yuan, K. Importance of CD44 in the Proliferation and Mineralization of Periodontal Ligament Cells. *J. Periodontal Res.* **2014**, *49*, 827–835. [CrossRef]
22. Bozic, D.; Grgurevic, L.; Erjavec, I.; Brkljacic, J.; Orlic, I.; Razdorov, G.; Grgurevic, I.; Vukicevic, S.; Plancak, D. The Proteome and Gene Expression Profile of Cementoblastic Cells Treated by Bone Morphogenetic Protein-7 in Vitro. *J. Clin. Periodontol.* **2012**, *39*, 80–90. [CrossRef]
23. Shimabukuro, Y.; Terashima, H.; Takedachi, M.; Maeda, K.; Nakamura, T.; Sawada, K.; Kobashi, M.; Awata, T.; Oohara, H.; Kawahara, T.; et al. Fibroblast Growth Factor-2 Stimulates Directed Migration of Periodontal Ligament Cells via PI3K/AKT Signaling and CD44/Hyaluronan Interaction. *J. Cell. Physiol.* **2011**, *226*, 809–821. [CrossRef] [PubMed]
24. Al-Rekabi, Z.; Fura, A.M.; Juhlin, I.; Yassin, A.; Popowics, T.E.; Sniadecki, N.J. Hyaluronan-CD44 Interactions Mediate Contractility and Migration in Periodontal Ligament Cells. *Cell Adhes. Migr.* **2019**, *13*, 139–151. [CrossRef] [PubMed]
25. Fujioka-Kobayashi, M.; Müller, H.-D.; Mueller, A.; Lussi, A.; Sculean, A.; Schmidlin, P.R.; Miron, R.J. In Vitro Effects of Hyaluronic Acid on Human Periodontal Ligament Cells. *BMC Oral Health* **2017**, *17*, 44. [CrossRef] [PubMed]
26. Asparuhova, M.B.; Kiryak, D.; Eliezer, M.; Mihov, D.; Sculean, A. Activity of Two Hyaluronan Preparations on Primary Human Oral Fibroblasts. *J. Periodontal Res.* **2019**, *54*, 33–45. [CrossRef]
27. Asparuhova, M.B.; Chappuis, V.; Stähli, A.; Buser, D.; Sculean, A. Role of Hyaluronan in Regulating Self-Renewal and Osteogenic Differentiation of Mesenchymal Stromal Cells and Pre-Osteoblasts. *Clin. Oral Investig.* **2020**, *24*, 3923–3937. [CrossRef]
28. Fujioka-Kobayashi, M.; Schaller, B.; Kobayashi, E.; Hernandez, M.; Zhang, Y.; Miron, R. Hyaluronic Acid Gel-Based Scaffolds as Potential Carrier for Growth Factors: An In Vitro Bioassay on Its Osteogenic Potential. *J. Clin. Med.* **2016**, *5*, 112. [CrossRef]
29. Zou, L.; Zou, X.; Chen, L.; Li, H.; Mygind, T.; Kassem, M.; Bünger, C. Effect of Hyaluronan on Osteogenic Differentiation of Porcine Bone Marrow Stromal Cells in Vitro. *J. Orthop. Res.* **2008**, *26*, 713–720. [CrossRef]
30. Kim, J.-J.; Song, H.Y.; Ben Amara, H.; Kyung-Rim, K.; Koo, K.-T. Hyaluronic Acid Improves Bone Formation in Extraction Sockets With Chronic Pathology: A Pilot Study in Dogs. *J. Periodontol.* **2016**, *87*, 790–795. [CrossRef]
31. Kim, J.; Ben Amara, H.; Park, J.; Kim, S.; Kim, T.; Seol, Y.; Lee, Y.; Ku, Y.; Rhyu, I.; Koo, K. Biomodification of Compromised Extraction Sockets Using Hyaluronic Acid and RhBMP-2: An Experimental Study in Dogs. *J. Periodontol.* **2019**, *90*, 416–424. [CrossRef] [PubMed]
32. Shirakata, Y.; Nakamura, T.; Kawakami, Y.; Imafuji, T.; Shinohara, Y.; Noguchi, K.; Sculean, A. Healing of Buccal Gingival Recessions Following Treatment with Coronally Advanced Flap Alone or Combined with a Cross-linked Hyaluronic Acid Gel. An Experimental Study in Dogs. *J. Clin. Periodontol.* **2021**, *48*, 570–580. [CrossRef]

33. Shirakata, Y.; Imafuji, T.; Nakamura, T.; Kawakami, Y.; Shinohara, Y.; Noguchi, K.; Pilloni, A.; Sculean, A. Periodontal Wound Healing/Regeneration of Two-Wall Intrabony Defects Following Reconstructive Surgery with Cross-Linked Hyaluronic Acid-Gel with or without a Collagen Matrix: A Preclinical Study in Dogs. *Quintessence Int. Berl. Ger. 1985* **2021**, *52*, 308–316. [CrossRef]
34. Bertl, K.; Bruckmann, C.; Isberg, P.-E.; Klinge, B.; Gotfredsen, K.; Stavropoulos, A. Hyaluronan in Non-Surgical and Surgical Periodontal Therapy: A Systematic Review. *J. Clin. Periodontol.* **2015**, *42*, 236–246. [CrossRef]
35. Eliezer, M.; Imber, J.-C.; Sculean, A.; Pandis, N.; Teich, S. Hyaluronic Acid as Adjunctive to Non-Surgical and Surgical Periodontal Therapy: A Systematic Review and Meta-Analysis. *Clin. Oral Investig.* **2019**, *23*, 3423–3435. [CrossRef]
36. Pilloni, A.; Rojas, M.A.; Marini, L.; Russo, P.; Shirakata, Y.; Sculean, A.; Iacono, R. Healing of Intrabony Defects Following Regenerative Surgery by Means of Single-Flap Approach in Conjunction with Either Hyaluronic Acid or an Enamel Matrix Derivative: A 24-Month Randomized Controlled Clinical Trial. *Clin. Oral Investig.* **2021**, *25*, 5095–5107. [CrossRef] [PubMed]
37. Miron, R.J.; Sculean, A.; Cochran, D.L.; Froum, S.; Zucchelli, G.; Nemcovsky, C.; Donos, N.; Lyngstadaas, S.P.; Deschner, J.; Dard, M.; et al. Twenty Years of Enamel Matrix Derivative: The Past, the Present and the Future. *J. Clin. Periodontol.* **2016**, *43*, 668–683. [CrossRef] [PubMed]
38. O'Leary, T.J.; Drake, R.B.; Naylor, J.E. The Plaque Control Record. *J. Periodontol.* **1972**, *43*, 38. [CrossRef]
39. Cortellini, P.; Prato, G.P.; Tonetti, M.S. The Simplified Papilla Preservation Flap. A Novel Surgical Approach for the Management of Soft Tissues in Regenerative Procedures. *Int. J. Periodontics Restor. Dent.* **1999**, *19*, 589–599.
40. Cortellini, P.; Prato, G.P.; Tonetti, M.S. The Modified Papilla Preservation Technique. A New Surgical Approach for Interproximal Regenerative Procedures. *J. Periodontol.* **1995**, *66*, 261–266. [CrossRef]
41. Wachtel, H.; Schenk, G.; Böhm, S.; Weng, D.; Zuhr, O.; Hürzeler, M.B. Microsurgical Access Flap and Enamel Matrix Derivative for the Treatment of Periodontal Intrabony Defects: A Controlled Clinical Study: Microsurgical Access Flap and EMD. *J. Clin. Periodontol.* **2003**, *30*, 496–504. [CrossRef]
42. Raines, A.L.; Sunwoo, M.; Gertzman, A.A.; Thacker, K.; Guldberg, R.E.; Schwartz, Z.; Boyan, B.D. Hyaluronic Acid Stimulates Neovascularization during the Regeneration of Bone Marrow after Ablation. *J. Biomed. Mater. Res. A* **2011**, *96A*, 575–583. [CrossRef]
43. Takeda, K.; Sakai, N.; Shiba, H.; Nagahara, T.; Fujita, T.; Kajiya, M.; Iwata, T.; Matsuda, S.; Kawahara, K.; Kawaguchi, H.; et al. Characteristics of High-Molecular-Weight Hyaluronic Acid as a Brain-Derived Neurotrophic Factor Scaffold in Periodontal Tissue Regeneration. *Tissue Eng. Part A* **2011**, *17*, 955–967. [CrossRef] [PubMed]
44. Nevins, M.; Giannobile, W.V.; McGuire, M.K.; Kao, R.T.; Mellonig, J.T.; Hinrichs, J.E.; McAllister, B.S.; Murphy, K.S.; McClain, P.K.; Nevins, M.L.; et al. Platelet-Derived Growth Factor Stimulates Bone Fill and Rate of Attachment Level Gain: Results of a Large Multicenter Randomized Controlled Trial. *J. Periodontol.* **2005**, *76*, 2205–2215. [CrossRef] [PubMed]
45. Cochran, D.L.; Oh, T.-J.; Mills, M.P.; Clem, D.S.; McClain, P.K.; Schallhorn, R.A.; McGuire, M.K.; Scheyer, E.T.; Giannobile, W.V.; Reddy, M.S.; et al. A Randomized Clinical Trial Evaluating Rh-FGF-2/β-TCP in Periodontal Defects. *J. Dent. Res.* **2016**, *95*, 523–530. [CrossRef] [PubMed]
46. Matarasso, M.; Iorio-Siciliano, V.; Blasi, A.; Ramaglia, L.; Salvi, G.E.; Sculean, A. Enamel Matrix Derivative and Bone Grafts for Periodontal Regeneration of Intrabony Defects. A Systematic Review and Meta-Analysis. *Clin. Oral Investig.* **2015**, *19*, 1581–1593. [CrossRef]
47. Briguglio, F.; Briguglio, E.; Briguglio, R.; Cafiero, C.; Isola, G. Treatment of Infrabony Periodontal Defects Using a Resorbable Biopolymer of Hyaluronic Acid: A Randomized Clinical Trial. *Quintessence Int. Berl. Ger. 1985* **2013**, *44*, 231–240. [CrossRef]
48. Vanden Bogaerde, L. Treatment of Infrabony Periodontal Defects with Esterified Hyaluronic Acid: Clinical Report of 19 Consecutive Lesions. *Int. J. Periodontics Restor. Dent.* **2009**, *29*, 315–323.
49. Leonardis, D.D.; Paolantonio, M. Enamel Matrix Derivative, Alone or Associated With a Synthetic Bone Substitute, in the Treatment of 1- to 2-Wall Periodontal Defects. *J. Periodontol.* **2013**, *84*, 444–455. [CrossRef]
50. Nibali, L.; Sultan, D.; Arena, C.; Pelekos, G.; Lin, G.; Tonetti, M. Periodontal Infrabony Defects: Systematic Review of Healing by Defect Morphology Following Regenerative Surgery. *J. Clin. Periodontol.* **2021**, *48*, 101–114. [CrossRef]
51. Trombelli, L.; Farina, R.; Vecchiatini, R.; Maietti, E.; Simonelli, A. A Simplified Composite Outcome Measure to Assess the Effect of Periodontal Regenerative Treatment in Intraosseous Defects. *J. Periodontol.* **2020**, *91*, 723–731. [CrossRef] [PubMed]
52. Aimetti, M.; Fratini, A.; Manavella, V.; Giraudi, M.; Citterio, F.; Ferrarotti, F.; Mariani, G.M.; Cairo, F.; Baima, G.; Romano, F. Pocket Resolution in Regenerative Treatment of Intrabony Defects with Papilla Preservation Techniques: A Systematic Review and Meta-analysis of Randomized Clinical Trials. *J. Clin. Periodontol.* **2021**, *48*, 843–858. [CrossRef] [PubMed]

Article

Biomimetic Ceramic Composite: Characterization, Cell Response, and In Vivo Biocompatibility

Hung-Yang Lin [1], Yi-Jung Lu [2,†], Hsin-Hua Chou [3,4], Keng-Liang Ou [5,6,7,8,9,10], Bai-Hung Huang [5,11], Wen-Chien Lan [6], Takashi Saito [8], Yung-Chieh Cho [5,12], Yu-Hsin Ou [13,14], Tzu-Sen Yang [15,*] and Pei-Wen Peng [16,*]

[1] Department of Dentistry, Fu Jen Catholic University Hospital, Fu Jen Catholic University, New Taipei City 242, Taiwan; a00207@mail.fjuh.fju.edu.tw
[2] Division of Family and Operative Dentistry, Department of Dentistry, Taipei Medical University Hospital, Taipei 110, Taiwan; yi_jung2002@yahoo.com.tw
[3] School of Oral Hygiene, College of Oral Medicine, Taipei Medical University, Taipei 110, Taiwan; hhchou@tmu.edu.tw
[4] Dental Department of Wan-Fang Hospital, Taipei Medical University, Taipei 116, Taiwan
[5] Biomedical Technology R & D Center, China Medical University Hospital, Taichung 404, Taiwan; klou@tmu.edu.tw (K.-L.O.); u109312001@cmu.edu.tw (B.-H.H.); D204106003@tmu.edu.tw (Y.-C.C.)
[6] Department of Oral Hygiene Care, Ching Kuo Institute of Management and Health, Keelung 203, Taiwan; jameslan@ems.cku.edu.tw
[7] Department of Dentistry, Taipei Medical University-Shuang Ho Hospital, New Taipei City 235, Taiwan
[8] Division of Clinical Cariology and Endodontology, Department of Oral Rehabilitation, School of Dentistry, Health Sciences University of Hokkaido, Hokkaido 061-0293, Japan; t-saito@hoku-iryo-u.ac.jp
[9] 3D Global Biotech Inc., New Taipei City 221, Taiwan
[10] Taiwan Society of Blood Biomaterials, New Taipei City 221, Taiwan
[11] Graduate Institute of Dental Science, College of Dentistry, China Medical University, Taichung 404, Taiwan
[12] School of Dentistry, College of Oral Medicine, Taipei Medical University, Taipei 110, Taiwan
[13] Excelsior School, Arcadia, CA 91007, USA; jennifer30526@gmail.com
[14] Research Center for Biomedical Devices and Prototyping Production, Taipei Medical University, Taipei 110, Taiwan
[15] Graduate Institute of Biomedical Optomechatronics, College of Biomedical Engineering, Taipei Medical University, Taipei 110, Taiwan
[16] School of Dental Technology, College of Oral Medicine, Taipei Medical University, Taipei 110, Taiwan
* Correspondence: tsyang@tmu.edu.tw (T.-S.Y.); apon@tmu.edu.tw (P.-W.P.)
† Co-first author: Yi-Jung Lu.

Citation: Lin, H.-Y.; Lu, Y.-J.; Chou, H.-H.; Ou, K.-L.; Huang, B.-H.; Lan, W.-C.; Saito, T.; Cho, Y.-C.; Ou, Y.-H.; Yang, T.-S.; et al. Biomimetic Ceramic Composite: Characterization, Cell Response, and In Vivo Biocompatibility. *Materials* 2021, 14, 7374. https://doi.org/10.3390/ma14237374

Academic Editors: Tobias Tauböck and Matej Par

Received: 15 October 2021
Accepted: 29 November 2021
Published: 1 December 2021

Publisher's Note: MDPI stays neutral with regard to jurisdictional claims in published maps and institutional affiliations.

Copyright: © 2021 by the authors. Licensee MDPI, Basel, Switzerland. This article is an open access article distributed under the terms and conditions of the Creative Commons Attribution (CC BY) license (https://creativecommons.org/licenses/by/4.0/).

Abstract: The present study aimed to synthesize biphasic calcium phosphate ceramics (CaPs) composed of β-tricalcium phosphate (β-TCP) and hydroxyapatite (HAp) from the propagated *Scleractinian* coral and dicalcium phosphate anhydrous using a solid-state reaction followed by heat treatment at a temperature of 1100 °C for 1 h to 7 days. The as-prepared coral and coral-derived biphasic CaPs samples were characterized through scanning electron microscopy, X-ray diffractometry, Fourier transform infrared spectroscopy, and Raman spectroscopy. The cell response of the biphasic CaPs was evaluated by in vitro cytotoxicity assessment using mouse fibroblast (L929) cells. The bilateral femoral defect rabbit model was used to assess the early local reaction of the coral-derived biphasic CaPs bone graft on tissue. The results confirmed that the co-existence of β-TCP and HAp was formed at 1100 °C for 1 h. The ratio of HA/β-TCP increased as the heat-treatment time increased. The coral-derived biphasic CaPs comprising 61% HAp and 39% β-TCP (defined as HT-3) were not cytotoxic. Furthermore, no significant differences in local tissue reaction were observed between the HT-3 sample and autogenous bone. Therefore, the synthesized coral-derived biphasic CaPs is a candidate for bone grafting due to its good biocompatibility.

Keywords: bioactive materials; bioceramics; biocompatibility; composites; calcium phosphate

1. Introduction

Coral exoskeletons possess unique interconnected porous architecture including tubular cavities ranging from 100 to 250 mm in length, similar to human bones and teeth, and have consequently attracted attention in orthopedics and maxillofacial surgery [1,2]. The presence of macropores greater than 150–500 μm in diameter facilitates nutrient diffusion, enables osteogenesis, and enhances bone formation [3]. Coral exoskeletons are crystals composed of calcium carbonate in the form of aragonite or calcite; their resorption rate as a bone growth rate is too fast to allow sufficient bone ingrowth, which limits clinical applications [4–6].

Hydroxyapatite (HAp, $Ca_{10}(PO4)_6(OH)_2$) is more similar to bone and teeth and shows lower solubility than calcium carbonates in the body fluids [7–9]. Therefore, biomimetic synthesis methods have been explored to include HAp and other CaP with the calcium carbonate microstructure with the interconnected macroporosity [10–14]. Both in vitro and in vivo studies have demonstrated that the bioceramics from these biogenic sources have dual functions of osteoconduction and osteoinduction [13,14].

The co-existence of HAp and β-tricalcium phosphate (β-TCP, $Ca_3(PO_4)_2$) was often observed under the routes to synthesize the pure HAp or β-TCP [15]. β-TCP has a comparable chemical composition to HAp; however, it is considered resorbable in vivo due to its faster release of Ca^{2+} and PO_4^{3-} ions when exposed to physiological fluids [16,17]. As bone graft substitutes, mixtures of the stable HAp with Ca/P ratio of 1.67 and the soluble β-TCP with Ca/P ratio of 1.5 have demonstrated higher efficacy for degradation rate to match the new bone formation rate when compared with single-phase HAP or β-TCP [18–20].

Hydrothermal conversion is the preferred method for the production of coral-derived HAp. However, this synthetic process is inappropriate for commercial manufacturing because this process is slow, has complicated pH adjustments, and limited β-TCP batch size [21–23]. Under the solid-state reaction route, a homogeneous mixture of calcium and phosphate precursors can be mixed in water at room temperature to synthesize a biphasic CP with a controlled ratio of HAp to β-TCP using various heat-treatment temperatures and durations [21,24]. Highly crystalline and biphasic CaPs synthesized were suitable for mass production, high reproducibility, and low processing cost [25].

The present study aimed to synthesize highly crystalline biphasic CaPs from propagated *Scleractinian* coral [26] and dicalcium phosphate anhydrous (DCPA, $CaHPO_4$) using a solid-state reaction followed by heat treatment. The influence was investigated of the heat-treated durations on the formation of the crystalline CaPs and their respective cytotoxicity properties. The rabbit bone defect model was used to assess the potential of coral-derived CPs as bone grafting.

2. Materials and Methods

2.1. As-Prepared Coral Granules

DCPA was used as a phosphate precursor. The propagated *Scleractinian* coral was purchased (Popeye Marine Biotechnology Limited, New Taipei City, Taiwan). Organic substances were removed using a self-developed cleaning process, and the coral was manually crushed and sieved using a 595 μm-mesh sieve. These deproteinized coral granules were washed thoroughly using the demineralized water, neutralized with phosphate-buffered saline (PBS), and sterilized in an autoclave. A commercial coral calcium powder, hereafter denoted as SMP-44 (Biomed herbal, Taichung, Taiwan), was used for comparison.

2.2. Coral-Derived Biphasic CaPs

The as-prepared coral granules and DCPA with the Ca to P (designated as Ca/P) molar ratio of 1.50 were mixed in demineralized water and homogenized in a brushless stirrer at a rotation speed of 450 rpm for 4 h. After filtering and drying, the mixture was heat-treated at 1100 °C in the high-temperature furnace (JH-4, Kingtech Sciencetific, Taipei, Taiwan) for a specified duration. The samples herein were referred to as HT-0.1, HT-3, and HT-7 when the heat treatment was for 1 h, 3 days, and 7 days.

2.3. Surface and Microstructure Analysis

Crystallinity analysis and phase identification of the heat-treated samples were performed using X-ray diffractometry (XRD; Rigaku 2200, Tokyo, Japan) operated with Cu Kα radiation operated at 50 kV and 250 mA. Crystalline phases were identified by comparing the database from the Joint Committee on Powder Diffraction Standards (JCPDS) [27]. The mass fraction was semi-quantified using the area of the peaks from β-TCP (0210) and the sum of the area of the peaks from HA (211) and β-TCP (0210). The chemical bonding information of the samples was characterized via Fourier-transform infrared spectroscopy (FTIR; Perkin-Elmer Spectrum 100, Shelton, CT, USA) with a spectral resolution of 4 cm^{-1}. Surface morphology of the samples was examined using scanning electron microscopy (SEM; JEOL-6500F, Tokyo, Japan) using an accelerating voltage of 20 kV [28,29]. Raman spectra were recorded at room temperature using Raman and a scanning near-field, optical microscope equipped with a 633 nm excitation laser source (Horiba HR800, Protrustech Co., Ltd., Taipei, Taiwan) [24,30].

2.4. In Vitro Cytotoxicity Evaluation

The mouse fibroblast cell line (L929 RM60091, Bioresource Collection, and Research Center, Hsinchu, Taiwan) was adopted in this experiment according to ISO 10993-5 specification [31]. The cells were seeded in culture dishes at a density of 5×10^4 cells per 100 μL in α-Minimum Essential Medium (MEM; Level Biotechnology, New Taipei City, Taiwan). Cells from passage 2 were harvested at 80% confluence and used for further 3-[4,5-dimethylthiazol-2-yl]-2,5-diphenyltetrazolim bromide (MTT) assay. The extracts of the investigated samples were placed in an orbital shaker maintained at 37 °C for 24 h with a mass to volume extraction ratio of 0.2 g/mL, which was followed by filtering and sealing in sterile bottles. L929 cells at a density of 1×10^4 cells/well were cultured in MEM and seeded on the 24-well culture plates. After obtaining a confluent monolayer, the medium was replaced by 0.1 mL sample extracts and incubated for 24 h at 37 °C in an atmosphere of 5% CO_2 (*n* = 3). Subsequently, a 10 μL MTT assay kit (R&D system, Minneapolis, MN, USA) was added to each well and incubated for 2 h. The optical density (OD) value of each plate was read at 570 nm using a microplate reader (ELx800, BioTek, Winooski, VT, USA). The cell viability is expressed as shown in Equation (1):

$$\text{Viability rate (\%)} = \frac{OD_{HT} - OD_b}{OD_{nc} - OD_b} \times 100\% \qquad (1)$$

where HT represents the measured OD of the samples, and nc and b represent the measured ODs of the negative control (NC) and the blank. The culture medium with 10% (*v/v*) dimethyl sulfoxide (DMSO) and an extract from high-density polyethylene were used as positive control (PC) and NC [32].

For qualitative evaluation, L929 cells at a density of 5×10^4 cells/well were seeded on the 24-well culture plates and incubated for 24 h at 37 °C in an atmosphere of 5% CO_2. The original culture medium was replaced by 0.5 mL sample extracts and incubated for 24 h at 37 °C in an atmosphere of 5% CO_2 (*n* = 3). Then, the cells were stained with the neutral red solution (Merck Taiwan, Taipei, Taiwan). Changes in cell morphology, cell lysis, and membrane integrity were observed using the inverted fluorescence microscope (FV1000/IX81, Olympus, Tokyo, Japan) under different magnifications.

2.5. A Pilot Study of the Rabbit Model for In Vivo Biocompatibility Assessment

A pilot study was carried out at Master Laboratory Co., Ltd. (Hsinchu, Taiwan) according to the standard of ISO 10993-6:2016 "Biological evaluation of medical devices—Part 6: Tests for local effects after implantation". Five New Zealand white rabbits weighing 2.8–3 kg (Livestock Research Institute, Tainan, Taiwan) were used to assess in vivo local tissue reaction using a bone defect model at the distal femur where autogenous bone was used for comparison. Before implantation, rabbits were sedated by intramuscular injection of tiletamine–zolazepam (10 mg/kg, Virbac Taiwan, Taipei, Taiwan) and xylazine hydrochloride (10 mg/kg, Bayer Taiwan, Taipei, Taiwan); then, they were anesthetized with isoflurane under aseptic conditions. Bone defects were created with a diameter of 4 mm and depth of 5 mm in the bilateral condyle of the femur using a motorized drill (Frios Unit S, Dentsply Sirona Taiwan, New Taipei City, Figure 1a). The bone graft sample was implanted into the defect site on the right femur, and the autogenous bone was implanted into the left femur (Figure 1b) by a well-trained veterinary. The rabbit was euthanized with carbon dioxide 4 weeks post-operatively. Bone blocks were collected from the adjacent and bottom regions of original bone defects for histological analysis. The bone blocks were fixed in 10% neutral-buffered formalin, dehydrated using ethanol, embedded in methyl methacrylate (MMA), sectioned into 4 to 5 μm-thick slices, stained with hematoxylin and eosin (H&E), and observed via using the Aperio CS pathology scanner (Leica Biosystems, Buffalo Grove, IL, USA).

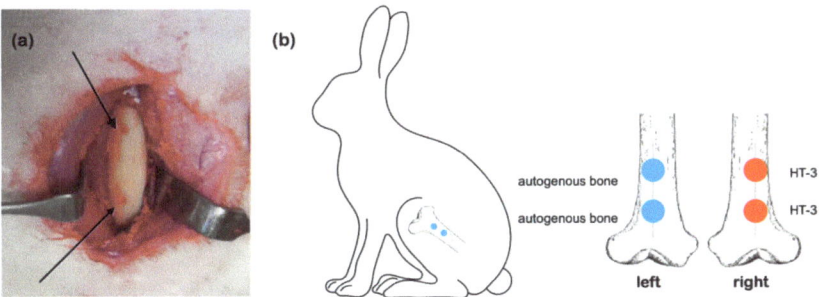

Figure 1. (**a**) Surgical model and (**b**) a schematic implantation design in the right distal femur of rabbit.

2.6. Statistical Analysis

The SPSS statistic software (Version 19.0., SPSS Inc., Chicago, IL, USA) was used to analyze the experimental data. The difference between groups was determined by one-way analysis of variance followed by Tukey's HSD post hoc test. A *p* value of less than 0.05 was considered as statistically significant.

3. Results

3.1. Characterizations of the Propagated Coral Granules

The SEM image in Figure 2a reveals the as-prepared coral granules with interconnected pore sizes of 100 to 250 μm. A polycrystalline and fibrous morphology was observed at the highly magnified image; as shown in Figure 2b–d, the SMP-44 sample of commercial coral calcium powder did not have the interconnected pores.

Figure 3a shows that the as-prepared coral exhibited a fully crystalline, single phase of aragonite (JCPDS 01-076-0606), which is similar to the literature [4]. The diffractogram of the SMP-44 sample shown in Figure 3b showed crystalline peaks corresponding to $(Ca,Mg)CO_3$, but the peaks slightly shifted to larger diffraction angles (JCPDS 00-005-0622).

FTIR spectra revealed that the as-prepared coral exhibited the characteristic bands for aragonite at 712 (in-plane bending mode, $v4$), 854 (out of plane bending mode, $v2$), 1082 (symmetric stretching mode $v1$), and 1472 (asymmetric stretching, $v3$) cm^{-1}, as shown in Figure 4a. The Raman spectroscopy confirmed the phases present, as shown in Figure 4b. The characteristic Raman symmetric stretching band ($v1$) at 1087.6 cm^{-1} and the in-plane bending mode ($v4$) at 705.8 cm^{-1} of the Raman spectra also indicate that the as-prepared coral granules were aragonite.

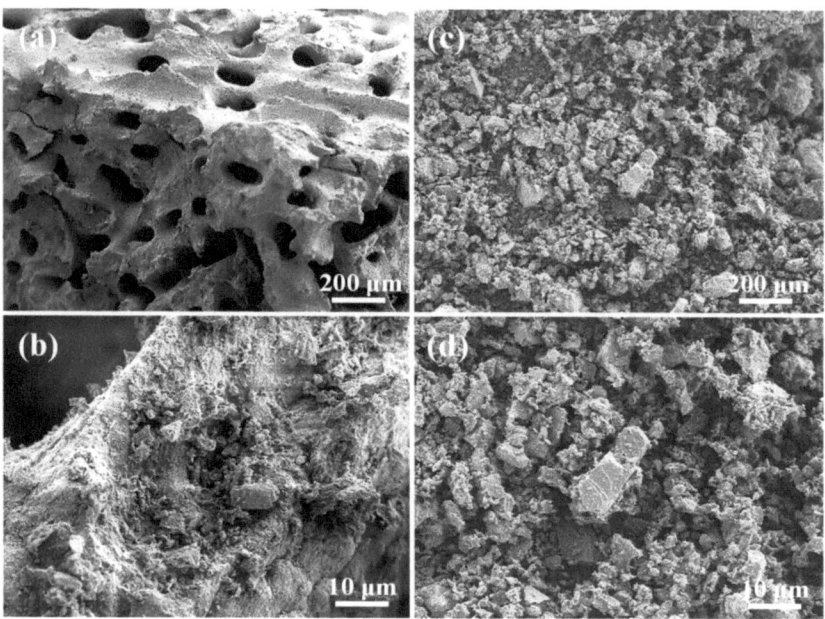

Figure 2. Comparisons between as-prepared coral and commercial coral calcium powder samples. (**a**) SEM image of the as-prepared coral, (**b**) a higher magnification SEM image of the as-prepared coral, (**c**) SEM image of the commercial coral calcium sample, and (**d**) a higher magnification SEM image of the commercial coral calcium sample.

3.2. Effects of Heat-Treated Duration on the Formation of Biphasic CaPs

Figure 5 shows the XRD patterns of the samples prepared by the heat treatment of as-prepared coral and DCPA at various durations. The diffraction spectra of all heat-treated samples were similar. A complete decomposition of $CaCO_3$ took place; these diffraction peaks were absent after heat treatment. Only β-TCP (JCPDS 00-009-0169) and HAp (JCPDS 01-084-1998) phases were observed for all heat-treated samples. The diffraction intensity of the peaks corresponding to β-TCP decreased with longer heat-treated time. Table 1 presents the results of semi-quantification of the crystalline phases in biphasic CaPs, showing that longer heat treatment created more HAp.

Table 1. Relative phase content (%) in biphasic CaP synthesized for varying durations.

Phase	HT-0.1	HT-3	HT-7
HAp	39.8	61.0	62.6
β-TCP	60.2	39.0	37.4

Figure 6 shows the effect of heat treatment duration on the FTIR spectra. The spectra of all heat-treated samples were similar, having the hydroxyl (OH^-) and phosphate (PO_4^{3-}) groups characteristic of CaPs. The sharp peak at 3643 cm^{-1}, belonging to the stretching vibration motion of the OH^- groups, was observed for all heat-treated samples. The bond regions of PO_4^{3-} groups at 611–540 cm^{-1} and 1139–944 cm^{-1} were also observed in all heat-treated samples, as were the asymmetrical stretching bands ν3 (1139 cm^{-1}), the symmetric stretching bands ν1 (944 cm^{-1}), and the bending bands ν4 (611, 586 and 543 cm^{-1}), which were related to PO_4^{3-} groups of β-TCP [13,29]. All samples also presented the asymmetrical P-O stretching mode (ν3) at 1086 cm^{-1}, the asymmetrical bending modes (ν4) at 600 and 568 cm^{-1}, and the symmetric stretching modes at 962 (ν1) and 474 (ν2) cm^{-1}, which are distinguishable peaks in the PO_4^{3-} groups of crystalline HAp [18]. Two peaks were exhibited at 3544 cm^{-1} and 3571 cm^{-1} that represented the stretching vibrations of the OH$^-$ groups in crystalline HAp. A trace of the symmetric stretching vibration of the HPO$_4^{2-}$ groups at 877 cm^{-1} appeared in all heat-treated samples.

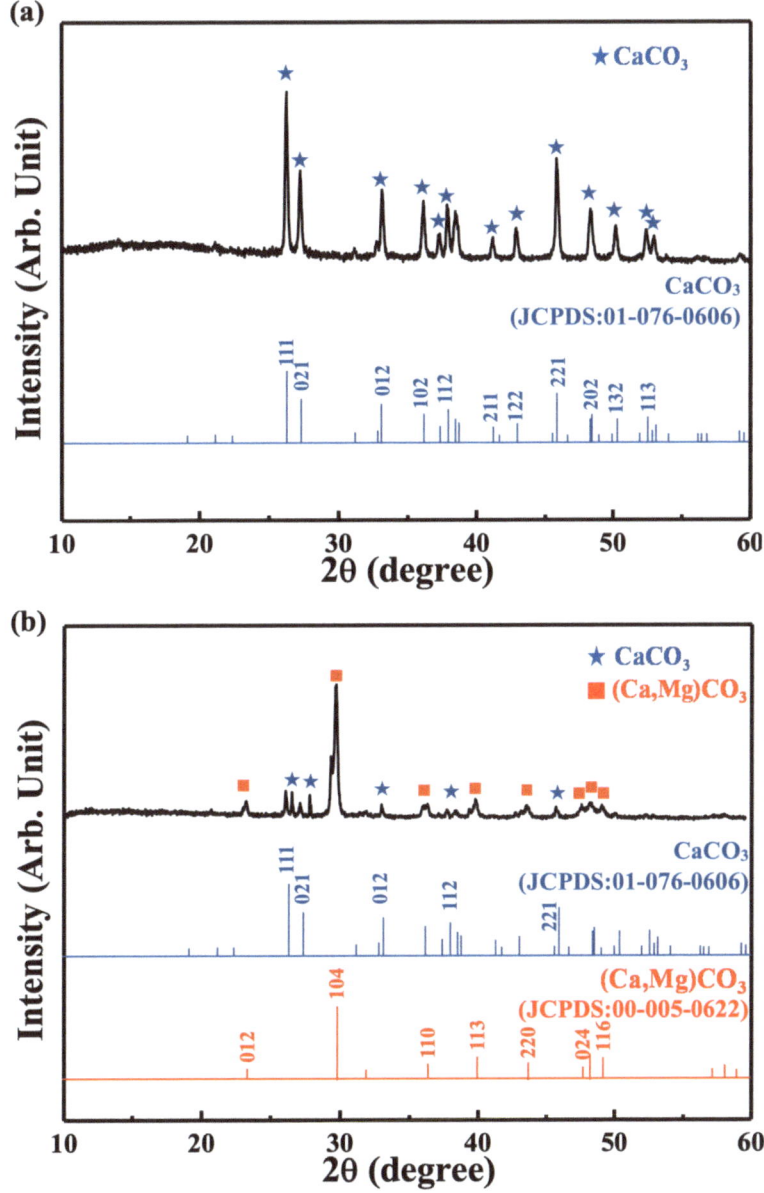

Figure 3. XRD pattern of (**a**) the as-prepared coral and (**b**) the commercial coral calcium sample.

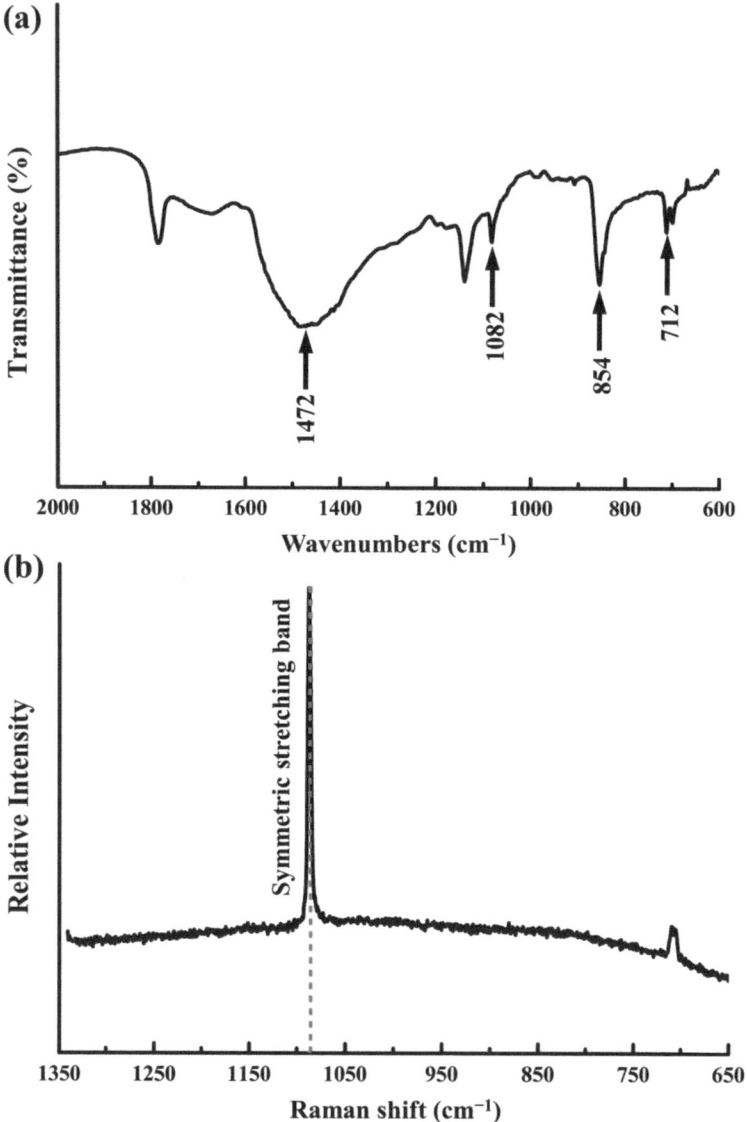

Figure 4. (a) FTIR and (b) Raman spectra of the as-prepared coral sample.

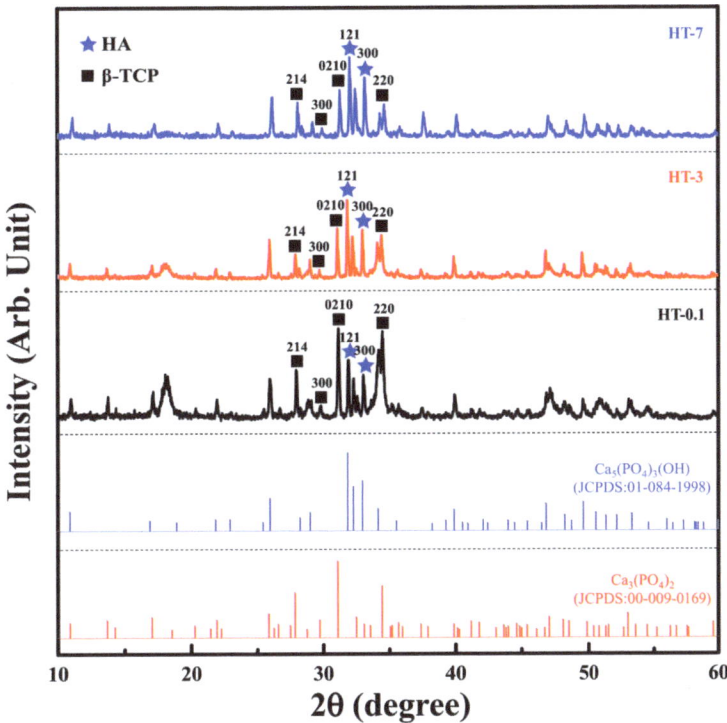

Figure 5. XRD patterns of the samples prepared by the heat treatment of as-prepared coral and DCPA at various durations, including reference spectra for JCPDS 01-084-1988 and 00-009-0169.

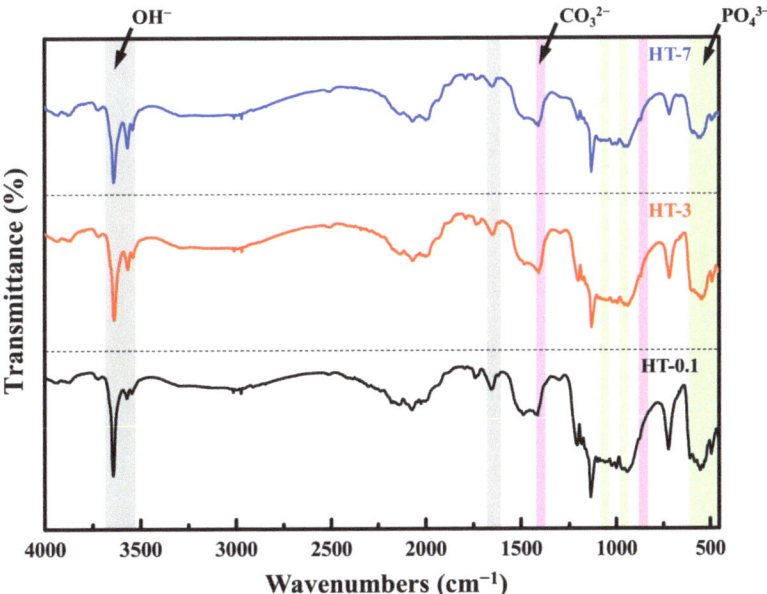

Figure 6. FTIR spectra of the samples prepared by the heat treatment of various durations of as-prepared coral and DCPA. Green areas indicate the bond regions of PO_4^{3-} groups; gray areas indicate the bond regions of OH^- groups; pink areas indicate the bond regions of CO_3^{2-} groups.

Figure 7 has the Raman spectra of the samples prepared by calcination of the as-prepared coral and DCPA at different durations in the fingerprint region (900–1100 cm^{-1}). All heat-treated samples exhibited the main vibration modes associated with PO_4^{3-} groups: a wideband representing symmetric P-O stretching mode (ν1) located in 930–990 cm^{-1}, and the asymmetric stretching (ν3) was at 1030–1080 cm^{-1}. Some differences were observed in the range between 950 and 980 cm^{-1}. The HT-0.1 sample displayed major peaks at 950, 972, and 1048 cm^{-1} that could be assigned to the vibrational internal modes of PO_4^{3-} groups of β-TCP. Some peaks at 964 and 1088 cm^{-1} were observed in the HT-0.1 sample, indicating the formation of HAp. With the increasing heat-treated time, the peak intensity at 964 cm^{-1} increased, and that at 1048 cm^{-1} decreased.

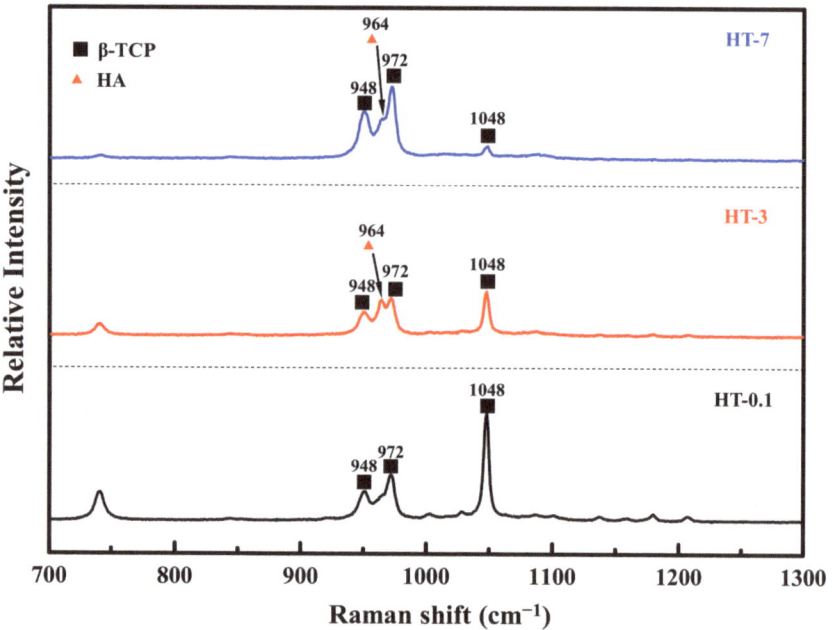

Figure 7. Raman spectra of the investigated samples prepared by the heat treatment of as-prepared coral and DCPA at different durations.

3.3. In Vitro Cytotoxicity Response

Based on the above investigations, the HT-3 sample was selected for the study of the heat-treatment conditions discussed in the in vitro cytotoxicity response and in vivo bone defect test. Table 2 has the OD values of cells that were co-cultured with extracts from the HT-3 sample for 24 h. No significant differences in cell viability were observed among the blank, NC, HT-3, and 50% HT-3 groups, and the cell viabilities were higher than those of the PC group. The cell viabilities of both 50% and 100% extracts of the HT-3 sample were higher than 70%, indicating that the HT-3 sample possessed no acute cytotoxic potential. Figure 8 shows the cell morphology variations of the tested samples. Cells had long spindle shapes with good density in the blank and NC groups (Figure 8a,b), whereas round cells and nearly destruction of the cell layers were observed in the PC group (Figure 8c). The HT-3 group (Figure 8d) had a cell morphology and density similar to the blank and NC groups.

Table 2. The results of MTT assay for evaluation of cell viability.

	OD$_{570\ nm}$	Viability (%)	Cell Lysis (%)
Blank	0.986 ± 0.002	100	0
NC	0.984 ± 0.003	100	0
PC	0.098 ± 0.001	10	90
HT-3	0.929 ± 0.035	94	6
50% HT-3	0.963 ± 0.028	98	2

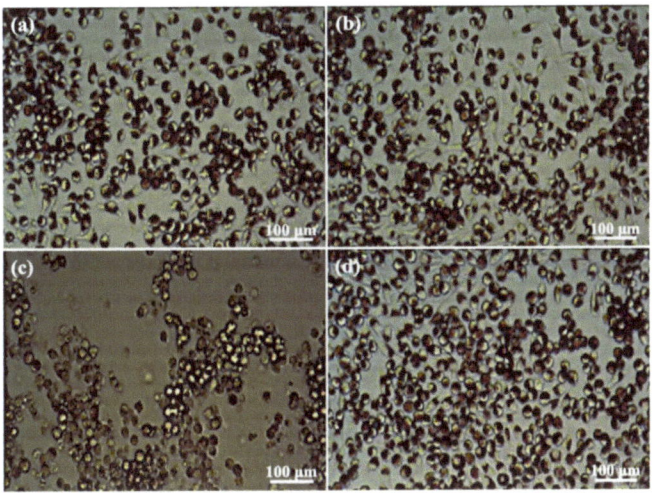

Figure 8. Optical images of the cell morphologies on the investigated samples: (**a**) blank, (**b**) negative control, (**c**) positive control, and (**d**) HT-3.

3.4. Bone Tissue Reaction Features

Figure 9 shows the histological images of the defects. No abnormal behavior or wound infection were found in the two graft materials, the coral samples, or autogenous bone, 4 weeks after the bone graft. Lymphocytes in the HT-3 group were higher than the autogenous bone group, although no significant differences were observed in the overall tissue reaction between the two materials (Table 3). These findings indicate that the HT-3 samples were considered non-responsive to tissues compared to autogenous bone samples after 4 weeks of implantation. The investigated HT-3 sample had no adverse effect on the tissue response and was comparable to that of the autogenous bone sample.

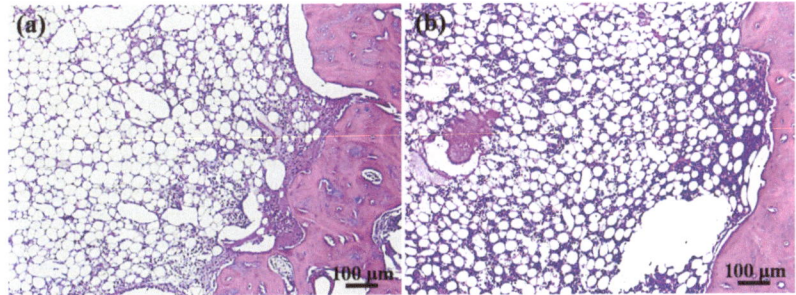

Figure 9. H & E staining results of the investigated samples after 4 weeks of implantation: (**a**) control and (**b**) HT-3.

Table 3. Histological findings and irritant-ranking scores according to ISO 10993-6: 2016.

	HT-3 (n = 10)	Autogenous Bone Group (n = 10)	p Value
Polymorphonuclea	1.2 ± 1.8	0.8 ± 1.1	0.68
Lymphocytes	4.4 ± 0.9	3.6 ± 0.9	0.20
Plasma cells	0	0	
Macrophages	0	0	
Giant cells	0	0	
Necrosis	0	0	
Neovascularization	0	0	
Fibrosis	0.6 ± 0.5	0.6 ± 0.5	>0.05
Fatty infiltrate	2.2 ± 0.5	2.4 ± 0.9	0.67

4. Discussion

Coral-derived CaP composed of HAp and β-TCP were obtained after heat treating the mixture of the propagated *Scleractinian* coral and DCPA at 1100 °C for as little as 1 h. The DCPA suspension with NaOH, NaF, or NaCl was transformed into HAp or TCP after hydrolysis for seven days; however, this process was slow and had complicated pH adjustments [21,22]. Furthermore, traditional porous-forming technologies were hard and relatively rare to realize the complex structures, which were structurally similar to human bone and tooth enamel [10,11]. The present study used the propagated *Scleractinian* coral as a calcium precursor and provided the calcium carbonate microstructure with the inter-connected microporosity.

β-TCP may be formed at 800 °C through the dehydration of DCPA to β-$Ca_2P_2O_7$ and decomposition of $CaCO_3$ to CaO [21,24]. Rhee et al. [25] synthesized mixtures of HAp and β-TCP using mechanochemical treatments. No noticeable changes of the initial two powders, $Ca_2P_2O_7$ and $CaCO_3$, were observed after 8 h of milling, but the surface area increased, which caused the speculation. Similar results were observed in the present study. A dominant component of the raw material used, the propagated *Scleractinian* coral, was aragonite $CaCO_3$. The heat treatment for 1 h resulted in the complete decomposition of the starting materials into only two crystalline phases: HA and β-TCP. The relative amount of β-TCP was higher than HAp after 1 h of heat treatment and decreased with longer heat treatment, implying that β-TCP was easily formed. The structural similarity between the starting materials and β-TCP facilitated the transformation from $CaCO_3$ into β-TCP [16]. In additional, the Ca/P ratio of the starting materials also provided favorable conditions for β-TCP formation [17].

The relatively greater proportion of HAp after increasing heat treatments may result from the instability of β-TCP [20]. With higher heated temperature, slight hydration and a slight deficit of hydration water in DCPA facilitated to form HAp, as shown in the following reaction [25]:

$$4Ca_3(PO_4)_4 + 2H_2O \rightarrow Ca_{10}(PO_4)_6(OH)_2(\text{surface}) + 2CaHPO_4.$$

The diffraction peaks of CaP-based mixtures were overlapped because of the similar chemical structures. Therefore, the vibrational spectroscopies helped characterize vibrations for these materials' amorphous species or crystalline phases [15,29]. The FTIR and Raman spectra results were in good agreement with the XRD findings in the present study. The presence of the Raman (Figure 5) OH^- peaks at 3645 cm^{-1} may result from the adsorbed moisture due to the highly hygroscopic property of phosphates and the appearance of the ambient air atmosphere during heat treatment [24]. The OH^- peak decreased with longer heat treatment, suggesting that the conversion of β-TCP to HA was formed due to the existence of water and the supply of additional hydroxyl groups [25].

The mixture heat-treated for 3 days (HT-3) contained 61% HAp and 39% β-TCP, and there was no significant difference in HAp/β-TCP ratio between HT-3 and HT-7 samples (Table 2). DMSO was used as a positive control because it was cytotoxic. High-density polyethylene was used as a negative control to clear the background response of the cells [32]. These results and the MTT assay demonstrated that heat treatment for 3 days was sufficient to generate biphasic CaP with the stable phase composition and excellent biocompatibility properties [16]. The results of qualitative visual cellular characterization shown in Figure 7 from the cytotoxicity test agreed with those of MTT assay analysis regarding the toxicity of the cells. The cell layer reactivity that resulted from biomaterial extracts can be scored from 0 (no reactivity) to 4 (severe reactivity) [31]. L929 cells exposed to HT-3 showed no reactivity (Grade 1) and were considered cytocompatible.

The pilot study of rabbit femur defects revealed that the HT-3 sample did not induce an irritant response 4 weeks after rabbit femur bone implantation. It had the same biological effect (without adverse effect) as the autogenous bone sample. Therefore, it is believed that the coral-derived biphasic CaPs comprising 61% HAp and 39% β-TCP (i.e., HT-3 sample) not only exhibited good biocompatibility but also possessed the potential to facilitate bone tissue regeneration. More studies should be performed to validate the present findings.

5. Conclusions

Biphasic calcium phosphates containing β-TCP and HAp were synthesized by stirring solutions of propagated coral and dicalcium phosphate anhydrous followed by heat treatments at 1100 °C for 1 h to 7 days. The synthesized biphasic CaPs samples displayed no cytotoxic effects. Implantation in bone defects in the rabbit model did not induce locally adverse tissue reactions and had new bone formation. The coral-derived biphasic CaPs with 61% HAp and 39% β-TCP sample are a useful bone graft substitute for bone defect treatment applications.

Author Contributions: Writing-original draft, H.-Y.L. and Y.-J.L.; Funding acquisition, H.-Y.L.; Investigation, Y.-J.L. and H.-H.C.; Conceptualization, H.-H.C. and K.-L.O.; Data curation, Y.-C.C. and Y.-H.O.; Methodology, B.-H.H. and W.-C.L.; Resources, Y.-C.C. and Y.-H.O.; Validation, B.-H.H. and W.-C.L.; Supervision, K.-L.O. and T.S.; Project administration, T.S.; Visualization, T.-S.Y. and P.-W.P.; Writing-review & editing, T.-S.Y. and P.-W.P. All authors have read and agreed to the published version of the manuscript.

Funding: The authors would like to thank the Fu Jen Catholic University Hospital, Taiwan for financially supporting this research under contract No. PL-202108024-V.

Institutional Review Board Statement: The protocols for animal experiments were reviewed and approved by the Institutional Animal Care and Use Committee for Master Laboratory Co., Ltd. under a project identification code of MSA-202001-20-T08.

Informed Consent Statement: Not applicable.

Data Availability Statement: Data are contained within the article.

Conflicts of Interest: The authors declare no conflict of interest.

References

1. Shi, Y.; Pan, T.; Zhu, W.; Yan, C.; Xia, Z. Artificial bone scaffolds of coral imitation prepared by selective laser sintering. *J. Mech. Behav. Biomed. Mater.* **2020**, *104*, 103664. [CrossRef] [PubMed]
2. Matuda, Y.; Okamura, T.; Tabata, H.; Yasui, K.; Tatsumura, M.; Kobayashi, N.; Nishikawa, T.; Hashimoto, Y. Periodontal Regeneration Using Cultured Coral Scaffolds in Class II Furcation Defects in Dogs. *J. Hard Tissue Biol.* **2019**, *28*, 329–334. [CrossRef]
3. Takeuchi, A.; Tsuge, T.; Kikuchi, M. Preparation of porous β-tricalcium phosphate using starfish-derived calcium carbonate as a precursor. *Ceram. Int.* **2016**, *42*, 15376–15382. [CrossRef]
4. Chou, J.; Hao, J.; Ben-Nissan, B.; Milthorpe, B.; Otsuka, M. Coral exoskeletons as a precursor material for the development of a calcium phosphate drug delivery system for bone tissue engineering. *Biol. Pharm. Bull.* **2013**, *36*, 1662–1665. [CrossRef]
5. He, F.; Zhang, J.; Yang, F.; Zhu, J.; Tian, X.; Chen, X. In vitro degradation and cell response of calcium carbonate composite ceramic in comparison with other synthetic bone substitute materials. *Mater. Sci. Eng. C* **2015**, *50*, 257–265. [CrossRef]
6. Neto, A.S.; Ferreira, J.M.F. Synthetic and Marine-Derived Porous Scaffolds for Bone Tissue Engineering. *Materials* **2018**, *11*, 1702. [CrossRef]

7. Cui, W.; Song, Q.; Su, H.; Yang, Z.; Yang, R.; Li, N.; Zhang, X. Synergistic effects of Mg-substitution and particle size of chicken eggshells on hydrothermal synthesis of biphasic calcium phosphate nanocrystals. *J. Mater. Sci. Technol.* **2020**, *36*, 27–36. [CrossRef]
8. Eliaz, N.; Metoki, N. Calcium phosphate bioceramics: A review of their history, structure, properties, coating technologies and biomedical applications. *Materials* **2017**, *10*, 334. [CrossRef]
9. Bohner, M.; Santoni, B.L.G.; Döbelin, N. β-tricalcium phosphate for bone substitution: Synthesis and properties. *Acta Biomater.* **2020**, *113*, 23–41. [CrossRef]
10. Roy, D.M.; Linnehan, S.K. Hydroxyapatite formed from coral skeletal carbonate by hydrothermal exchange. *Nature* **1974**, *247*, 220–222. [CrossRef]
11. Jinawath, S.; Polchai, D.; Yoshimura, M. Low-temperature, hydrothermal transformation of aragonite to hydroxyapatite. *Mater. Sci. Eng. C* **2002**, *22*, 35–39. [CrossRef]
12. Rosa Cegla, R.-N.; Macha, I.J.; Ben-Nissan, B.; Grossin, D.; Heness, G.; Chung, R.-J. Comparative study of conversion of coral with ammonium dihydrogen phosphate and orthophosphoric acid to produce calcium phosphates. *J. Aust. Ceram. Soc.* **2014**, *50*, 154–161.
13. Nandi, S.K.; Kundu, B.; Mukherjee, J.; Mahato, A.; Datta, S.; Balla, V.K. Converted marine coral hydroxyapatite implants with growth factors: In vivo bone regeneration. *Mater. Sci. Eng. C* **2015**, *49*, 816–823. [CrossRef]
14. Karacan, I.; Ben-Nissan, B.; Sinutok, S. Marine-Based Calcium Phosphates from Hard Coral and Calcified Algae for Biomedical Applications. In *Marine-Derived Biomaterials for Tissue Engineering Applications*; Springer: Berlin/Heidelberg, Germany, 2019; pp. 137–153.
15. Vallet-Regi, M.; González-Calbet, J.M. Calcium phosphates as substitution of bone tissues. *Prog. Solid State Chem.* **2004**, *32*, 1–31. [CrossRef]
16. Webler, G.; Zapata, M.; Agra, L.; Barreto, E.; Silva, A.; Hickmann, J.; Fonseca, E. Characterization and evaluation of cytotoxicity of biphasic calcium phosphate synthesized by a solid state reaction route. *Curr. Appl. Phys.* **2014**, *14*, 876–880. [CrossRef]
17. Mirjalili, F.; Bagheshahi, S.; Aghaee, M. Synthesis and characterization of β-TCP/HA nanocomposite: Morphology and microstructure. *J. Thermoplast. Compos. Mater.* **2020**, *33*, 1292–1313. [CrossRef]
18. Roopavath, U.K.; Sah, M.K.; Panigrahi, B.B.; Rath, S.N. Mechanochemically synthesized phase stable and biocompatible β-tricalcium phosphate from avian eggshell for the development of tissue ingrowth system. *Ceram. Int.* **2019**, *45*, 12910–12919. [CrossRef]
19. Zhang, L.; Zhang, C.; Zhang, R.; Jiang, D.; Zhu, Q.; Wang, S. Extraction and characterization of HA/β-TCP biphasic calcium phosphate from marine fish. *Mater. Lett.* **2019**, *236*, 680–682. [CrossRef]
20. Ho, W.-F.; Hsu, H.-C.; Hsu, S.-K.; Hung, C.-W.; Wu, S.-C. Calcium phosphate bioceramics synthesized from eggshell powders through a solid state reaction. *Ceram. Int.* **2013**, *39*, 6467–6473. [CrossRef]
21. Laonapakul, T.; Sutthi, R.; Chaikool, P.; Talangkun, S.; Boonma, A.; Chindaprasirt, P. Calcium phosphate powders synthesized from $CaCO_3$ and CaO of natural origin using mechanical activation in different media combined with solid-state interaction. *Mater. Sci. Eng. C* **2021**, *118*, 111333. [CrossRef]
22. Guo, X.; Yan, H.; Zhao, S.; Li, Z.; Li, Y.; Liang, X. Effect of calcining temperature on particle size of hydroxyapatite synthesized by solid-state reaction at room temperature. *Adv. Powder Technol.* **2013**, *24*, 1034–1038. [CrossRef]
23. Hou, P.-J.; Lee, C.-Y.; Ou, K.-L.; Lan, W.-C.; Chuo, Y.-C.; Lin, H.-Y.; Chao, H.-W.; Huang, B.-H.; Saito, T.; Tsai, H.-Y. Calcium release from different toothpastes after the incorporation of tricalcium phosphate and amorphous calcium phosphate. *Appl. Sci.* **2021**, *11*, 1848. [CrossRef]
24. Choi, D.; Kumta, P.N. Mechano-chemical synthesis and characterization of nanostructured β-TCP powder. *Mater. Sci. Eng. C* **2007**, *27*, 377–381. [CrossRef]
25. Rhee, S.-H. Synthesis of hydroxyapatite via mechanochemical treatment. *Biomaterials* **2002**, *23*, 1147–1152. [CrossRef]
26. Barton, J.A.; Willis, B.L.; Hutson, K.S. Coral propagation: A review of techniques for ornamental trade and reef restoration. *Rev. Aquac.* **2017**, *9*, 238–256. [CrossRef]
27. Liu, S.; Chen, J.; Chen, T.; Zeng, Y. Fabrication of trabecular-like beta-tricalcium phosphate biomimetic scaffolds for bone tissue engineering. *Ceram. Int.* **2021**, *47*, 13187–13198. [CrossRef]
28. Ferreira, J.; Kannan, S. Phase transition mechanisms involved in the formation of structurally stable β-$Ca_3(PO_4)$2-α-Al_2O_3 composites. *J. Eur. Ceram. Soc.* **2017**, *37*, 2953–2963.
29. Sinusaite, L.; Grigoraviciute-Puroniene, I.; Popov, A.; Ishikawa, K.; Kareiva, A.; Zarkov, A. Controllable synthesis of tricalcium phosphate (TCP) polymorphs by wet precipitation: Effect of washing procedure. *Ceram. Int.* **2019**, *45*, 12423–12428. [CrossRef]
30. Aguiar, H.; Chiussi, S.; López-Álvarez, M.; González, P.; Serra, J. Structural characterization of bioceramics and mineralized tissues based on Raman and XRD techniques. *Ceram. Int.* **2018**, *44*, 495–504. [CrossRef]
31. Narayan, R. *Encyclopedia of Biomedical Engineering*; Elsevier: Amsterdam, The Netherlands, 2018.
32. Wallin, R.F.; Arscott, E. A practical guide to ISO 10993-5: Cytotoxicity. *Med. Device Diagn. Ind.* **1998**, *20*, 96–98.

Article

Long-Term Assessment of Contemporary Ion-Releasing Restorative Dental Materials

Danijela Marovic [1], Matej Par [1,*], Karlo Posavec [2], Ivana Marić [3], Dominik Štajdohar [1], Alen Muradbegović [4], Tobias T. Tauböck [5], Thomas Attin [5] and Zrinka Tarle [1]

1. Department of Endodontics and Restorative Dentistry, School of Dental Medicine, University of Zagreb, Gunduliceva 5, 10000 Zagreb, Croatia; marovic@sfzg.hr (D.M.); dstajdohar@sfzg.hr (D.Š.); tarle@sfzg.hr (Z.T.)
2. Private Dental Practice, Dr. Ivana Novaka 28, 40000 Čakovec, Croatia; karlo.posavec.ned@gmail.com
3. Private Dental Practice, Odranska 10, 10000 Zagreb, Croatia; ivana.maric912@gmail.com
4. Private Dental Practice, Malkočeva 3, 75000 Tuzla, Bosnia and Herzegovina; muradbegovic.alen@gmail.com
5. Clinic of Conservative and Preventive Dentistry, Center for Dental Medicine, University of Zurich, Plattenstrasse 11, 8032 Zurich, Switzerland; tobias.taubock@zzm.uzh.ch (T.T.T.); thomas.attin@zzm.uzh.ch (T.A.)
* Correspondence: mpar@sfzg.hr

Abstract: The objective was to evaluate new commercially available ion-releasing restorative materials and compare them to established anti-cariogenic materials. Four materials were tested: alkasite Cention (Ivoclar Vivadent) in self-cure or light-cure mode, giomer Beautifil II (Shofu), conventional glass-ionomer Fuji IX (GC), and resin composite Tetric EvoCeram (Ivoclar Vivadent) as a control. Flexural strength, flexural modulus, and Weibull modulus were measured one day, three months, and after three months with accelerated aging in ethanol. Water sorption and solubility were evaluated for up to one year. Degree of conversion was measured during 120 min for self-cured and light-cured Cention. In this study, Beautifil II was the ion-releasing material with the highest flexural strength and modulus and with the best resistance to aging. Alkasite Cention showed superior mechanical properties to Fuji IX. Weibull analysis showed that the glass-ionomer had the least reliable distribution of mechanical properties with the highest water sorption. The solubility of self-cured alkasite exceeded the permissible values according to ISO 4049. Degree of conversion of light-cured Cention was higher than in self-cure mode. The use of alkasite Cention is recommended only in the light-cure mode.

Keywords: flexural strength; modulus; water sorption; solubility; degree of conversion; alkasite; giomer; glass-ionomer; long-term

Citation: Marovic, D.; Par, M.; Posavec, K.; Marić, I.; Štajdohar, D.; Muradbegović, A.; Tauböck, T.T.; Attin, T.; Tarle, Z. Long-Term Assessment of Contemporary Ion-Releasing Restorative Dental Materials. *Materials* **2022**, *15*, 4042. https://doi.org/10.3390/ma15124042

Academic Editor: Gherlone Felice Enrico

Received: 13 April 2022
Accepted: 3 June 2022
Published: 7 June 2022

Publisher's Note: MDPI stays neutral with regard to jurisdictional claims in published maps and institutional affiliations.

Copyright: © 2022 by the authors. Licensee MDPI, Basel, Switzerland. This article is an open access article distributed under the terms and conditions of the Creative Commons Attribution (CC BY) license (https:// creativecommons.org/licenses/by/ 4.0/).

1. Introduction

In recent years, restorative dentistry has gradually shifted from "biocompatibility" to "bioactivity". With the advancement of minimally invasive dentistry, scientific interest in restorative materials with ion release also increases [1–5]. Restorative materials should possess anti-demineralizing and remineralizing properties to fight against caries while retaining their stability over time and resistance to occlusal load, thermal changes, and enzymatic influences in the oral cavity.

The term "bioactive dental restorative materials" is still a matter of debate. While some biomaterial scientists claim that a bioactive material should be able to form a hydroxyapatite precipitate at its surface [6], others discard this idea [7]. At the same time, they should create an active interface with biological tissue [8]. Glass-ionomers were the first dental restorative materials able to satisfy some of the properties desired for the bioactive restorative material [9,10]. Fluoride release is considered accountable for promoting biomineralization of mineral-depleted hydroxyapatite [6,11], while self-adhesion to tooth substrate enables their direct interaction with hard dental tissues [9]. Glass-ionomers are hydrophilic materials and need water for their setting reaction. Still, they are also sensitive to dehydration (leading to cracking of the material surface) [12] and excessive water uptake (leading to the dissolution of metal cations) [13]. The clinical applicability of glass-ionomers is limited to low-stress bearing areas because of poor flexural strength, toughness, and wear [14,15].

Aiming to improve the mechanical properties and durability of conventional glass-ionomers to the level of resin composites [16–18], a variety of ion-releasing materials based on fluoroaluminosilicate glass as a filler component has been made: cermets, fiber-reinforced glass-ionomers, resin-modified glass-ionomers, compomers, and giomers [14,19]. The coupling of functional fillers and the methacrylate matrix is desirable to allow quick and on-demand hardening of a material. Giomers typically contain a resin-based matrix and unique pre-reacted glass-ionomer (PRG) fillers, which have a conventional glass core with a surface glass-ionomer layer pretreated with polyalkenoate acid and a completed acid–base reaction. PRG fillers are afterward dehydrated and silanated to ensure copolymerization to the resin [10]. Besides PRG fillers, giomers contain conventional silanated macro- and micro-fillers. This approach seems to be highly successful in terms of the giomer's high fracture toughness and flexural strength [19,20]. The fluoride release depends on the material's water sorption after placement in the moist environment and is therefore significantly lower than in resin-modified glass-ionomers or compomers [10,19]. Their behavior is considered very similar to resin composite, and their clinical performance is satisfactory [21].

Recently, a new class of resin-based ion-releasing materials appeared on the market, named alkasite materials. The name is derived from their alkalizing properties due to the release of hydroxide (OH^-) ions. The only material in that class is produced by Ivoclar Vivadent (Schaan, Liechtenstein), whose composition was modified over time. Cention N (Ivoclar Vivadent) was the first material that appeared on the market in the hand-mix version. Cention (Ivoclar Vivadent) and Cention forte (Ivoclar Vivadent) are their successors in a capsulated version. According to the manufacturer, the composition of Cention is the same as that of Cention forte, the difference being in the application mode (Cention forte is recommended for use with a special adhesive system). Three main types of fillers are present: silanized inert barium aluminum silicate glass, calcium barium aluminum fluorosilicate glass similar to glass-ionomers, and calcium fluorosilicate glass or "alkasite" glass. Besides these components, the manufacturer states that Cention also contains ytterbium trifluoride and a prepolymerized filler termed Isofiller, similar to other materials from the same manufacturer. The liquid phase consists of dimethacrylates without any acidic groups that would impart self-adhesive properties [10]. Cention is a bulk-fill restorative material with photoinitiators and chemical catalysts enabling a dual-cure polymerization mechanism. This material releases Ca^{2+}, F^- and PO_4^{3-} ions in neutral and acidic conditions, leading to apatite formation on its surface [22,23]. A series of studies by Par et al. showed that Cention has an acid-neutralizing capability [4] and prevented demineralization of enamel [24] and dentine [25] when subjected to lactic acid over a prolonged period. Presently, this material is considered the only true commercially available bioactive composite [10,23]. Clinical studies are still lacking, as well as the investigations on the influence of mineral deposits at the surface of the restoration on proper oral hygiene maintenance and the antimicrobial action [26,27].

The release of ions or any other substances from a restorative material always raises concerns about the possible dissolution of functional filler particles. In the set material placed in an aqueous environment, this could create voids and facilitate water sorption, propagating further dissolution. Internal porosities lower the resistance of restoration to occlusal forces and facilitate their fracture [28]. A compromise between satisfactory mechanical properties and the ion-releasing benefits is needed. While mechanical properties of resin composites [18,29,30], glass-ionomers [14,15], and giomers [19,21] are sufficiently explored, studies focused on alkasite materials are scarce and mainly investigate the powder-liquid hand-mixed Cention N [31–34]. Besides the work of Par and co-workers [4,22,24,25] that focused on ion-releasing properties of Cention, a PubMed search of articles including the capsulated version of Cention resulted in finding only three papers studying fluoride release [35], wear behavior [36], or biologic effects on pulp cells [37]. The data about the long-term mechanical behavior of capsulated alkasite Cention used in either self-cure or light-cure mode is still lacking, especially considering the compositional modifications of the capsulated version in contrast to the predecessor Cention N.

This study was thus conducted to examine the long-term influence of water and aging on the mechanical properties of currently available ion-releasing materials. Six parameters were tested: flexural strength and modulus, Weibull modulus, degree of conversion, water sorption, and solubility. The null-hypotheses were: (I) there is no difference between different materials in any of the tested parameters, (II) for any given parameter, there is no difference between different time points, and (III) there is no difference between Cention when light-cured or self-cured in any of the tested parameters.

2. Materials and Methods

Four materials were tested in this study (Table 1), but with five testing groups, as one material, alkasite Cention, was tested in a light-cured (LC) and self-cured (SC) mode.

Table 1. The composition of the tested materials provided by the manufacturers.

Type	Product Name (Manufacturer)	Composition	Curing Mechanism
Alkasite	Cention (Ivoclar Vivadent)	Powder: inert barium alumino-boro-silicate glass, ytterbium fluoride, a calcium fluoro-alumino-silicate glass, and a reactive SiO_2-CaO-CaF_2-Na_2O glass Liquid: UDMA, aromatic aliphatic UDMA, DCP, and PEG-400-DMA Initiator system: hydroperoxide, Ivocerin, and acyl phosphine oxide Filler content: 58–59 vol%	Dual-cure
Giomer	Beautifil II (Shofu Dental GmbH)	Fillers: s-PRG (aluminofluoro-borosilicate glass), Al_2O_3 Resin: bis-GMA, TEGDMA Filler content: 69 vol%	Light-cure
Glass-ionomer	Fuji IX GP Fast (GC Europe)	Powder: fluoro-alumino-silicate glass Liquid: Polybasic carboxylic acid (copolymer of acrylic and maleic acid), tartaric acid, water	Self-cure
Composite (control)	Tetric EvoCeram (Ivoclar Vivadent)	Fillers: Barium glass filler, ytterbium fluoride, mixed oxide, prepolymers Resin: bis-GMA, UDMA, bis-EMA Filler content: 53–55 vol%	Light-cure

Abbreviations: Bis-GMA—bisphenol-A-glycidyldimethacrylate; TEGDMA—triethylene glycol dimethacrylate; UDMA—urethane dimethacrylate; bis-EMA—ethoxylated bisphenol A-dimethacrylate; s-PRG—surface-modified pre-reacted glass-ionomer fillers.

2.1. Study Protocol

Three tests were performed (degree of conversion, three-point bending, and water sorption), and six parameters were measured: flexural strength, flexural modulus, Weibull modulus, water sorption, solubility, and degree of conversion (only for Cention). The study design is depicted in Figure 1.

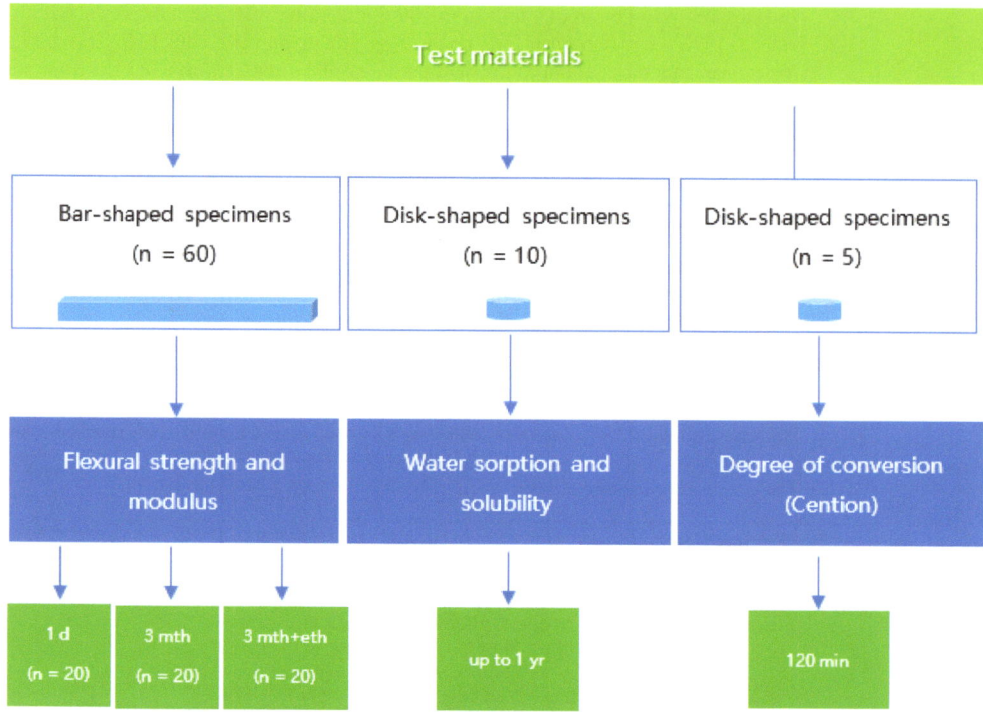

Figure 1. Flow chart of the study design.

2.2. Three-Point Bending Test

For the three-point bending test, bar-shaped specimens with dimensions 16 × 2 × 2 mm were made [24]. Unset materials were filled in a custom-made silicone mold (Elite HD+Putty, Zhermack, Badia Polesine, Italy) in excess, pressed with a polyethylene terephthalate (PET) foil, and a microscope cover glass and flash material was removed. Light-curing was performed for Cention LC, Beautifil II, and Tetric EvoCeram using Bluephase G2 (Ivoclar Vivadent) with 950 mW/cm^2 for 20 s, with three overlapping exposures on each side, making six irradiations in total. The radiant exitance of the curing unit was measured using a calibrated and NIST-referenced UV–Vis spectrophotometer (MARC; BlueLight Analytics, Halifax, NS, Canada) and amounted to 952 mW/cm^2 with peak intensities at 405 and 457 nm. Cention SC and Fuji IX were left to set at room temperature for 15 min in the dark. All specimens were then immersed in distilled water and stored at 37 °C in the dark.

Sixty specimens per group were subjected to a three-point bending test using a customized universal testing machine (Ultratester, Ultradent Products Inc., South Jordan, UT, USA). Twenty specimens in each group were tested after one day in distilled water. Another 20 specimens were tested after three months (90 d) in distilled water, while the remaining 20 specimens were tested after storage in distilled water for three months, followed by immersion in absolute ethanol for three days. Flexural strength and modulus were calculated [25]. The Weibull analysis (reliability analysis) was performed by plotting the function ln:

$$\ln(1/(1-P_f)) = m(\ln \sigma - \ln \sigma_\theta) \quad (1)$$

where P_f = probability of failure, m = Weibull modulus σ = strength at failure, and σ_θ = characteristic strength.

2.3. Water Sorption and Solubility

Ten disk-shaped specimens per material were made (2 mm high and 6 mm in diameter) in Teflon molds. The setting of the materials was performed similarly to the three-point bending test: Cention LC, Beautifil II, and Tetric EvoCeram were light-cured with the identical curing unit for 20 s on each side. At the same time, Cention SC and Fuji IX were left to set at room temperature for 15 min in the dark.

After initial drying in the desiccator, the specimens were weighted with an analytical scale (NBL 254 i, Adam Equipment, Milton Keynes, UK). The obtained values were designated as the initial mass of the specimen (m_1). Afterward, the specimens were individually placed in conical-shaped Eppendorf tubes with 4 mL of distilled water. They were stored for one year (365 days) at 37 °C in the dark. The mass of the specimens was weighted after 1, 7, 14, 90, 180, and 365 days ($m_2(t)$, t—time). After the immersion, the specimens were again dried in a desiccator. Their mass was regularly monitored until stable values (not differing from a previous measurement for more than 0.1 mg) were achieved. The final mass of the specimens after drying was marked as m_3.

Water sorption and solubility were calculated according to the formula provided by ISO 4049 [25]:

$$\text{water sorption} = (m_2(eq) - m_3) \text{ (g)} \tag{2}$$

$$\text{solubility} = m_1 - m_3 \text{ (g)} \tag{3}$$

where $m_2(eq)$ represents mass equilibrium.

2.4. Degree of Conversion

The degree of conversion was measured for alkasite Cention in self-cure or light-cure mode, using Fourier transform infrared (FTIR) spectrometer (Nicolet iS50, Thermo Fisher, Madison, NJ, USA) with an attenuated total reflectance (ATR) accessory. Cention capsules were mixed, and the material was extruded directly on the diamond ATR crystal using custom-made silicone molds at room temperature (22 ± 1 °C). The specimens (d = 6 mm, h = 1.5 mm) were covered with PET strips and left to self-cure or light-cured for 20 s using Bluephase G2. The curing unit was positioned perpendicularly and in direct contact with the composite specimen surface. FTIR spectra were continuously collected at a rate of 2 spectra per second for 120 min after the placement of the material or start of light-activated curing, with 4 scans and a resolution of 8 cm^{-1} [26]. Five specimens per experimental group were tested (n = 5).

The ratio between the peak heights of aliphatic (1638 cm^{-1}) and aromatic (1608 cm^{-1}) bands were used to calculate the degree of conversion (DC) for each spectrum for uncured and cured specimens. The degree of conversion was plotted against time.

$$\text{DC (\%)} = \left(1 - \frac{(1638 \text{ cm}^{-1}/1608 \text{ cm}^{-1}) \text{ after curing}}{(1638 \text{ cm}^{-1}/1608 \text{ cm}^{-1}) \text{ before curing}}\right) \times 100\% \tag{4}$$

2.5. Statistical Analysis

The normality of distribution was evaluated using Shapiro Wilk's test and the inspection of normal Q-Q diagrams. Since the data for flexural strength and modulus data violated the assumption of normality, the comparisons performed were statistically analyzed using the Kruskal-Wallis test with Bonferroni post-hoc adjustment. Weibull statistics were performed to examine the reliability of the materials. For water sorption and solubility, data were normally distributed, hence why the mixed-model ANOVA with Tukey and Bonferroni corrections (for independent and dependent observations, respectively) were used for statistical analysis. The degree of conversion data for Cention SC and Cention LC were normally distributed and compared using a t-test for independent observations. SPSS (version 20, IBM, Armonk, NY, USA) was used for the statistical analysis with the level of significance α = 0.05.

3. Results

Light-cured materials exhibited the highest flexural strength, followed by self-cured materials, in a decreasing manner: Tetric EvoCeram = Beautifil II > Cention LC > Cention SC > Fuji IX. Figure 2 shows that the flexural strength of the Cention LC was the highest after 1 day (104 ± 32 MPa) and after 3-month water exposure (99 ± 13 MPa), while significantly decreasing ($p = 0.003$) after ethanol immersion (84 ± 13 MPa). On the contrary, the same material showed a flexural strength increase when left to self-cure, so the 1-day values (62 ± 13 MPa) were significantly lower ($p < 0.001$) than values after 3-month water exposure (78 ± 16 MPa) and an additional ethanol immersion (87 ± 21 MPa, $p < 0.05$). Beautifil II demonstrated unexpectedly higher flexural strength values after 3-month water and ethanol exposure than after 3-month exposure to water only ($p = 0.032$).

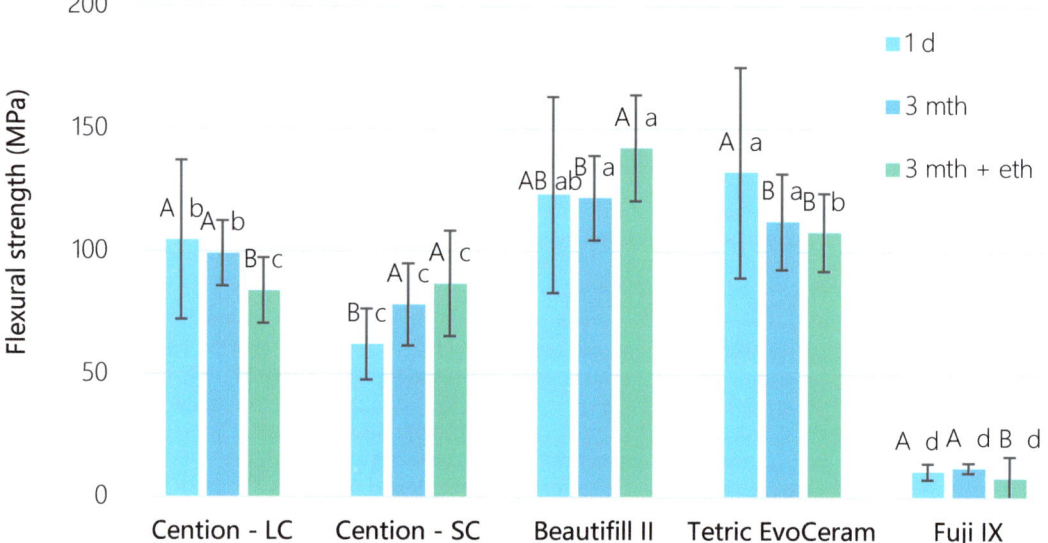

Figure 2. Flexural strength as a function of time for tested materials (mean values ± standard deviation, n = 20). Identical uppercase letters denote $p > 0.05$ for the same material between different time points; identical lowercase letters denote $p > 0.05$ between materials at the same time point.

A similar pattern was noted for the flexural modulus, as depicted in Figure 3. Cention LC demonstrated a significant drop down ($p < 0.001$) in modulus after 3 mth (4.2 ± 0.3 GPa) and 3 mth + eth (3.4 ± 0.3 GPa) groups compared to 1-day values (5.6 ± 1.7 GPa). The flexural modulus of Cention SC was significantly lower ($p = 0.001$) after 1 day (2.8 ± 0.5 GPa) than after 3 mth (3.3 ± 0.5 GPa) and 3 mth + eth (3.5 ± 0.6 GPa). Beautifil II showed a higher modulus ($p < 0.001$) after ethanol exposure (5.8 ± 0.5 Gpa) than after 3-month water exposure (5.4 ± 0.6 GPa). Fuji IX had the significantly lowest ($p < 0.001$) flexural strength (7.9–12.0 MPa) and the lowest flexural modulus (0.5–2.7 GPa).

Material reliability was calculated by the Weibull analysis (Figure 4). All light-cured groups showed high reliability and similarly narrow distribution of values, except after one day of water immersion. Unlike them, Cention SC had higher reliability with closely distributed values for one day. The Cention SC group demonstrated similar values, but these values were slightly lower compared to Cention LC. Fuji IX showed a wide distribution of data and, therefore, much lower reliability than other materials in this study.

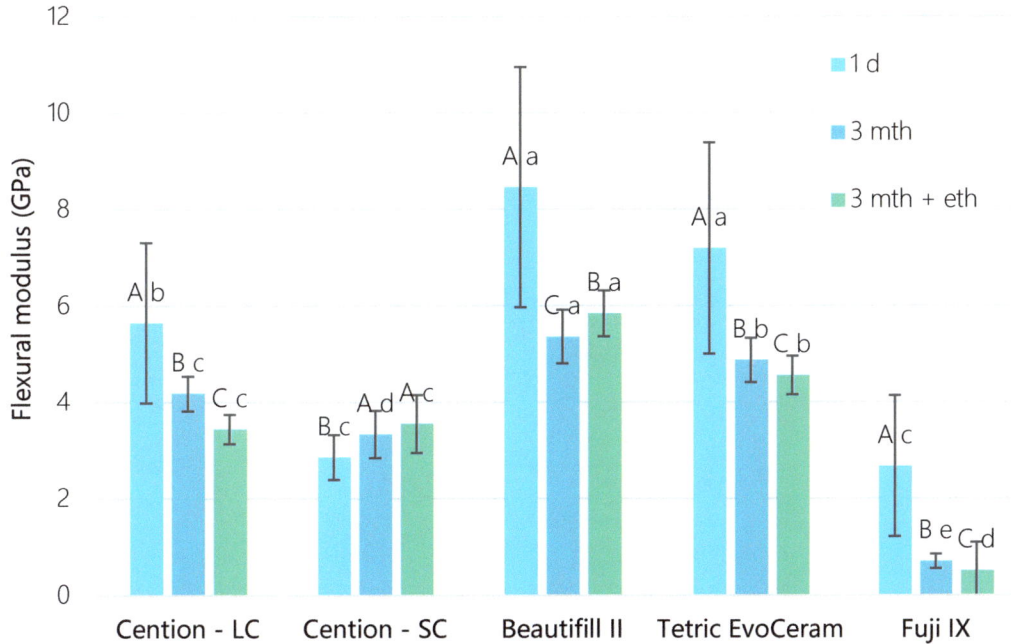

Figure 3. Flexural modulus as a function of time for tested materials (mean values ± standard deviation, n = 20). Identical uppercase letters denote $p > 0.05$ for the same material between different time points; identical lowercase letters denote $p > 0.05$ between materials at the same time point.

Figure 5 shows the results of water sorption and solubility. Fuji IX exhibited the highest water sorption (127.6 μg/mm^3), followed by Cention SC (73.6 μg/mm^3), Cention LC (40.5 μg/mm^3), while Tetric EvoCeram (31.6 μg/mm^3) and Beautifil II (30.4 μg/mm^3) had the lowest sorption ($p < 0.001$). The highest solubility was demonstrated by Cention SC (193.9 μg/mm^3), which was significantly reduced ($p < 0.001$) by photo-polymerization in Cention LC (21.9 μg/mm^3). Full water saturation was achieved after 90 days for Tetric EvoCeram and Cention LC, and after 180 days for Cention SC. After 365 days, a plateau of mass change was not reached for Beautifil II and Fuji IX, as visible from Figure 6. The highest mass gain for Fuji IX was accomplished during the first day (Figure 7), while the weight of Cention SC continuously dropped after the seventh day and continued falling for six months. At the 3-month point, Cention SC was the material with the lowest mass ($p = 0.001$–0.082), indicating mass loss. At the same time, Fuji IX had the highest mass ($p < 0.001$), while light-cured groups (Tetric EvoCeram, Beautifil II, and Cention LC) behaved statistically similarly.

The increase in the degree of conversion for Cention LC started immediately after activation of the curing unit, while for Cention SC, it started 11 min after mixing (Figure 8). The degree of conversion after 120 min was significantly higher ($p = 0.007$) for Cention LC (65.0 ± 2.1%) than for Cention SC (59.7 ± 2.5%).

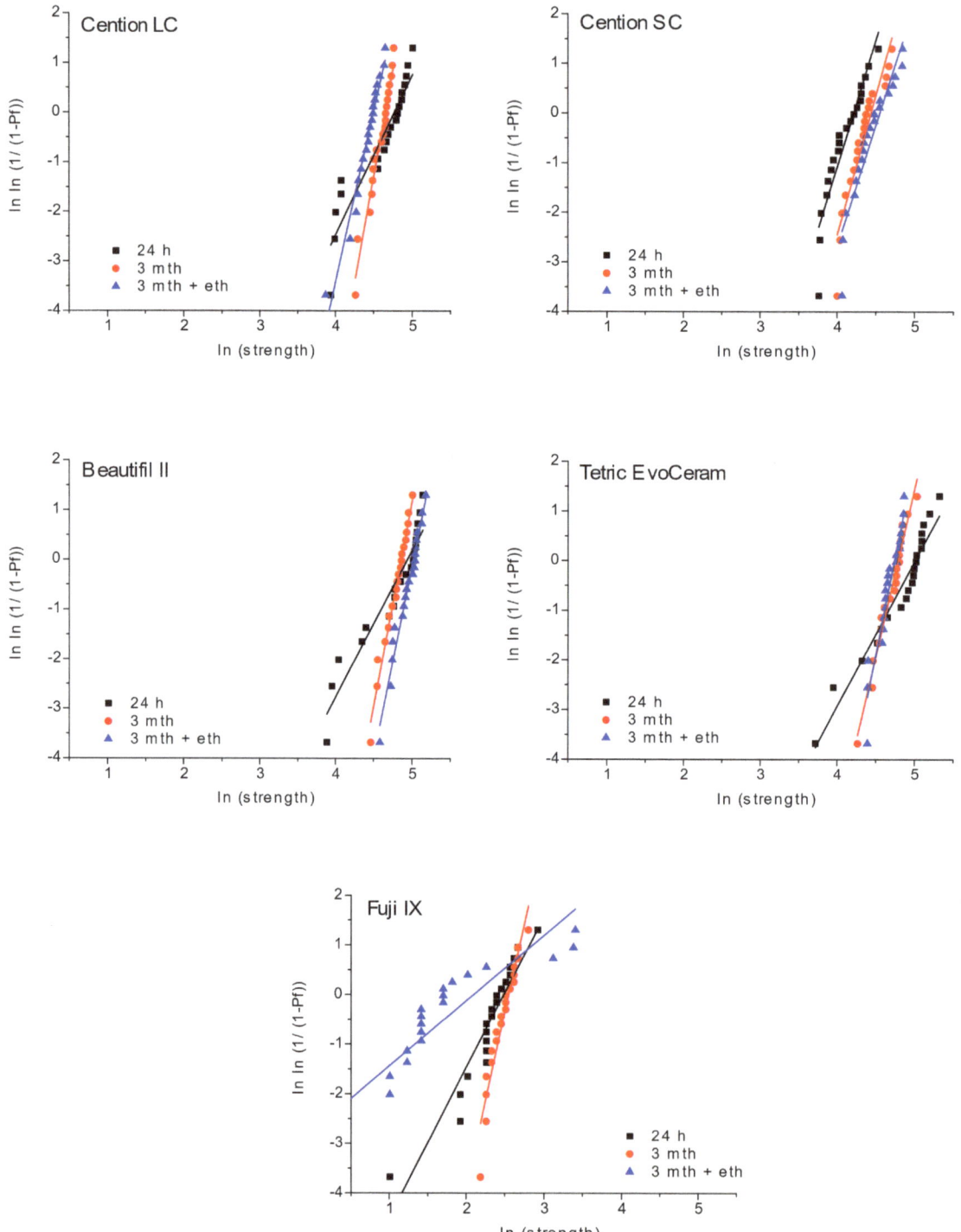

Figure 4. Weibull plots for tested materials and time points of measurement.

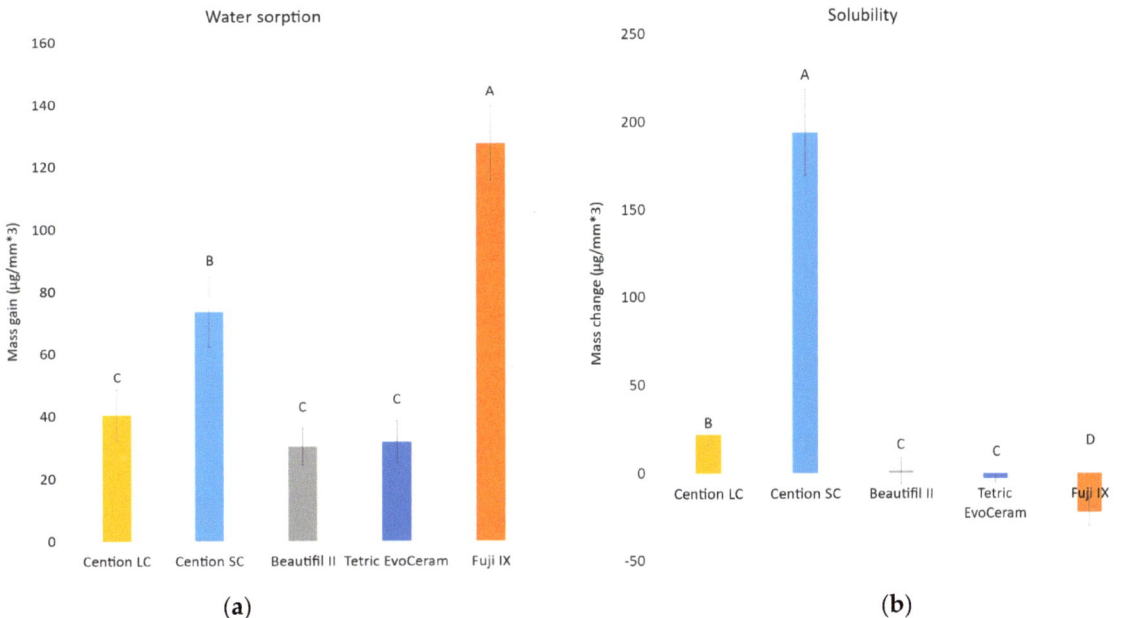

Figure 5. Water sorption (**a**) and solubility (**b**) for tested materials (mean values ± standard deviation, n = 10). Identical uppercase letters denote $p > 0.05$.

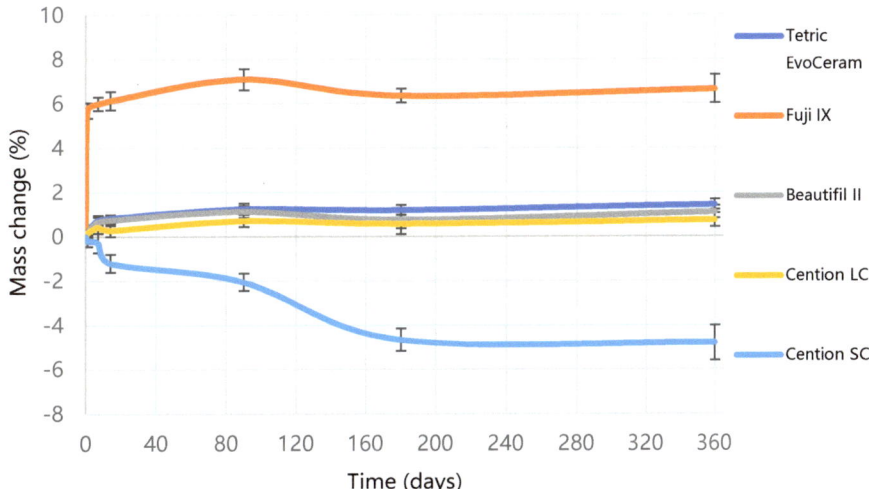

Figure 6. Mass change over one year of water immersion for tested materials. Error bars denote standard deviations.

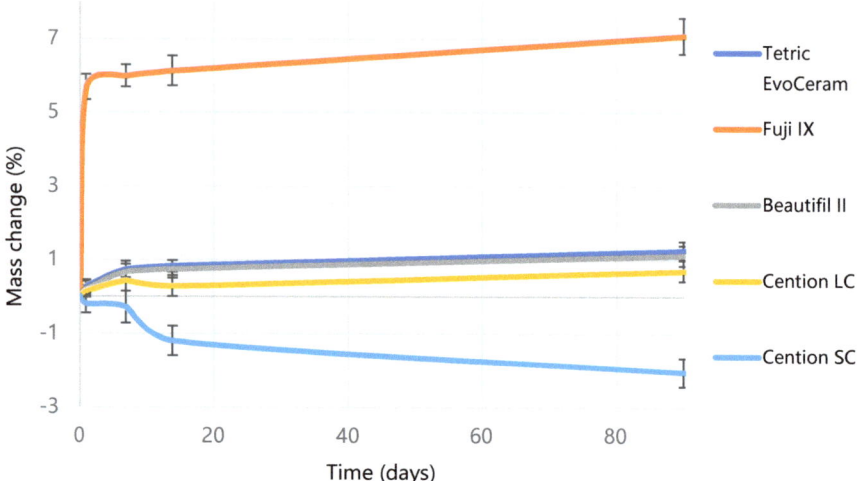

Figure 7. Mass changes over three months of water immersion. Error bars denote standard deviations.

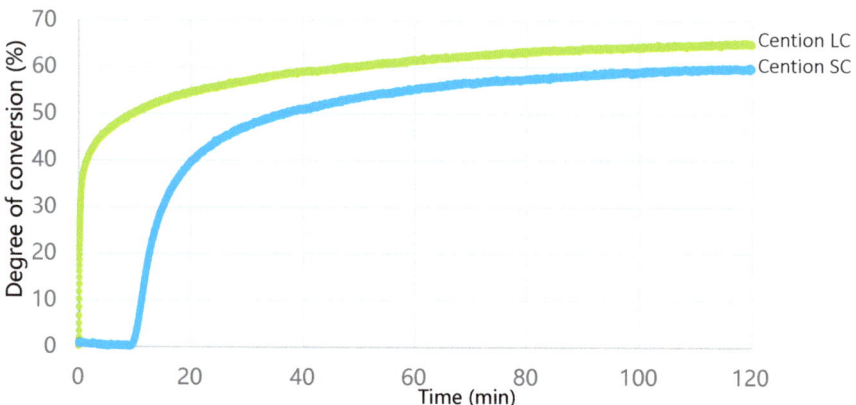

Figure 8. Degree of conversion as a function of time for Cention LC (green line) and Cention SC (blue line).

4. Discussion

This study examined the evolution of mechanical properties of ion-releasing materials over three months and after accelerated aging in ethanol, while water sorption and solubility were evaluated over one year. It was found that the flexural properties of the new bioactive composite Cention were higher than those of a high-viscosity glass-ionomer and lower than those of a conventional resin composite. When left to self-cure, this dual-cure bulk-fill material exhibited a slow increase in flexural strength and modulus as well as increased solubility. On the contrary, when light-cured, Cention showed slightly lower values than other light-cured materials in terms of mechanical properties and water sorption.

The distinct behavior of Cention in the self-cured and light-cured mode likely originated from different polymerization kinetics and resulting polymer networks. While the polymerization rate is the highest during light irradiation for the light-cured composites, redox polymerization in the self-curing modality has a delayed onset [38]. Ilie showed that initiation of polymerization of Cention N (hand-mixed) in a self-cure mode is lagging for 3.5 min after hand mixing, and that 11 min is needed to attain the same degree of conversion as in the light-cured mode [33]. However, our polymerization kinetics data on a capsulated Cention demonstrate the 11-min delay in initiating the polymerization of the self-cure mode, which prolongs the manufacturers' claimed working time from 2 to 11 min and extends the claimed setting time of 6.5 min. In this study, the polymerization reaction was monitored over two hours, and the self-cured Cention never reached the same degree of conversion as when light-cured. This is in accordance with two recent studies that found a significantly lower degree of conversion of self-cured vs. light-cured for a majority of tested dual-cured resin composites [34,38].

A delay in the polymerization activation of Cention SC led to a quick drop in mass of the water sorption specimens, indicating high solubility. The solubility of Cention was likely related to the dissolution of the functional fillers in an aqueous environment. It is necessary to underline that the present study diverged from the ISO 4049 recommendations for self-cured polymer materials (Class 1) in preparing the specimens [39]. While the ISO recommends the 60 min setting time, we opted for a more clinically relevant 15 min setting. The apparent instability of the self-cured specimens was reflected in the initially low flexural strength (62 ± 13 MPa) and modulus (2.8 ± 0.5 GPa) of the 1-day specimens. This is in contrast to the previous study on a predecessor material Cention N that allowed the 60 min setting and found much higher 1-day values (~100–120 MPa flexural strength and ~4–5 GPa modulus) [33]. The observed discrepancies are evidently related to the study design and the compositional modifications that had to be made for adjustment to the trituration of a capsulated Cention. The mass loss of Cention SC continued at a 3-month time point, reaching the equilibrium only after 6 months. However, mechanical properties improved over time despite the solubility.

This behavior could be explained by the fact that dense and highly cross-linked polymer network yields higher strength and modulus of a resin-based composite [40,41]. Even though no long-term measurements of degree of conversion were made, we can hypothesize that the gradual development of polymer cross-linking could have contributed to a delayed increase in flexural strength and modulus in Cention SC. At the same time, self-curing enabled uniform polymerization throughout the entire specimen, which apparently led to a close distribution of flexural strength values and practically no aging-induced change in reliability for Cention SC. Unfortunately, initial flexural strength and modulus values fall below the values recommended by ISO 4049, so using this material without the additional light-curing is not advised. Light-curing of the surface could act as an umbrella, protecting the deeper layers from the detrimental influence of water. However, the flexural strength and modulus gradient could cause an uneven distribution of forces and material fracture during the service life of the restoration. Further investigations in this field are necessary.

In the present study, accelerated aging in ethanol diminished flexural strength only in the light-cured Cention specimens, but not the self-cured. This phenomenon could be attributed to a significantly higher refractive index of alkaline fillers in contrast to conventional inert glass fillers [33]. Due to large filler/resin refractive index discrepancy, higher opacity of Cention in comparison to other bulk-fill composites is noted. Consequently, low light transmission is found, leading to only 13% light penetrating the material at 2 mm depth and 3% at a 4 mm level [33]. Considering that the photoinitiators in Cention are dibenzoyl germanium derivative and an acyl phosphine oxide, photoactivation in the violet part of the spectrum around 409 nm is optimal. Unfortunately, violet wavelengths reach shorter depths than blue due to exponential light attenuation. Higher opacity and lower light transmission could have led to an inhomogeneous polymer network with a decreasing cross-linking density. Such heterogeneous networks consist of highly crosslinked microgel agglomerates surrounded by less cross-linked polymer [41,42]. Ethanol as an organic solvent quickly penetrates the parts of the polymer network with fewer chemical cross-links, separates physical (hydrogen) bonds, and causes plasticization of the resin [43,44]. This degradation of the polymer network was probably reflected in the reduction in strength and modulus for light-cured Cention in the 3 mth + eth group. Contrary, self-cured Cention presumably achieved more uniform cross-linking throughout the entire thickness of the specimen. However, the polymerization reaction for Cention SC occurred at a much slower pace, which led to statistical difference in flexural strength and modulus between 1-day and both 3 mth and 3 mth + eth groups.

Similar to Cention SC, an unexpected rise of flexural strength and modulus was observed for giomer Beautifil II after artificial aging in ethanol. In the present study, the specimens were subjected to ethanol exposure to provoke maximum plasticization of the organic matrix and thus give the worst possible outcome of flexural properties [43,45]. Beautifil II has the highest filler volume in this study and, therefore, less organic matrix that could be susceptible to plasticization [19]. Still, this is not a complete explanation for the unusual behavior demonstrated after ethanol exposure. To the authors' knowledge, there are no studies that subjected Beautifil II to long-term water storage and ethanol after long-term water storage. However, its predecessor, Beautifil, was studied by Yap et al. [46]. They compared 30 days of water storage at 37 °C and 5000 thermal cycles varying between temperatures from 15 °C, 35 °C, and 45 °C. They found increased modulus and hardness after thermal cycling, explained by the post-cure polymerization due to heat exposure [46]. This explanation cannot be applied to the present study. Considering that the Beautifil II contains a traditional bis-GMA/TEGDMA matrix, we can only speculate that the ethanol-related increase can be associated with the unique S-PRG filler. However, the exact answer is still to be elucidated in future studies.

On the other hand, glass-ionomer Fuji IX was predictably the least reliable material in the study, with wide flexural strength data distribution, the highest water sorption, and generally lowest flexural properties. Such behavior is well described in the literature and can be attributed to high water sorption due to increased mobility of the sodium ion in the functional glass at room temperatures. Sodium is exchanged for hydrogen ions and causes hydrolytic instability and high solubility [14]. In addition to the hydrophilicity of glass-ionomers, internal porosities were identified as the origins of water accumulation, dissolution, and degradation of mechanical properties [14,15]. High water sorption of Fuji IX in the present study was thus expected and within the range of values described earlier [47–49]. Negative solubility for Fuji IX indicates incomplete water evaporation. Water was likely permanently bound during the cement's maturation as this hydrophilic material uses water in the setting process [50]. Similar behavior, but to a smaller extent, was noted for the reference composite material Tetric EvoCeram. These negative solubility values were reported in previous studies for the same material [18,51]. The literature describes that the water remained bound by the hydrogen bonds to the -OH groups in the methacrylates of the resin matrix [52].

According to ISO 4049, the maximum allowed water sorption for a polymer-based restorative material is 40 $\mu g/mm^3$ and 7.5 $\mu g/mm^3$ for solubility [39]. Both self-cured materials in this study overstepped these limits, even though ISO 4049 does not apply to conventional glass-ionomers. Cention SC showed the highest solubility (194 ± 24 $\mu g/mm^3$), while Fuji IX had the highest water sorption (127 ± 12 $\mu g/mm^3$). The insufficient curing could again explain the solubility of Cention SC compared to Cention LC. Water absorbed in partially polymerized specimens could cause leaching out of the unpolymerized monomers and, to a smaller extent, dissolution of functional fillers, loss of mass, and higher solubility [50]. The most significant weight loss of Cention SC specimens occurred during the first six months after water immersion but continued up to one year, as illustrated in Figures 6 and 7. Cention LC, on the other hand, behaves similarly to other light-cured materials. Thus, it is essential to reiterate that Cention should always be light-cured when placed in the oral cavity.

The hydrolytic deterioration of mechanical properties of polymer-based materials is significant [45,53], especially in ion releasing materials [54]. The time factor plays an important role in the diffusion of water or ethanol throughout the materials. The 24 h exposure to water proposed by ISO 4049 [39] seems insufficient to estimate the behavior of a material in a clinical setting [55,56]. Therefore, long-term studies such as the present one are necessary for evaluating ion-releasing materials. However, water sorption and solubility were not correlated to the decline of mechanical properties of all materials in this study. Previously, water sorption and flexural properties were related to filler type and amount, monomer composition, silanization and polymer crosslinking density [18,53,55]. The high filler ratio was the probable reason for the high long-term aging resistance of giomer Beautifil II, comparable or better than the inert reference material. On the other hand, high water sorption was likely the cause for the deterioration of the mechanical properties of glass-ionomer Fuji IX. New functional restorative material, alkasite Cention, showed a similar but slightly lower sorption and mechanical behavior pattern as an inert composite control, but only when light-cured. Considering the low ion-releasing ability of giomers [10,19], and poor mechanical properties of tested glass ionomer, it seems that alkasite Cention could be a viable ion-releasing alternative to conventional composite resins.

5. Conclusions

In conclusion, our results indicate that, when light-cured, Cention's mechanical and water sorption properties are satisfactory and better than the mechanical properties of a glass-ionomer tested here. Leaving the Cention to self-cure will cause lower polymerization of the material, high solubility, and poor mechanical properties immediately after placement. Alkasite Cention should be used only in the light-cure mode.

Author Contributions: Conceptualization, D.M. and M.P.; methodology, K.P., D.Š., I.M. and A.M.; software, K.P., M.P. and I.M.; validation, D.M. and M.P.; formal analysis, K.P. and M.P.; investigation, K.P., D.Š., I.M. and A.M.; resources, D.M., Z.T., T.T.T. and T.A.; data curation, K.P. and M.P.; writing—original draft preparation, D.M.; writing—review and editing, D.M., M.P., K.P., D.Š., I.M., A.M., Z.T., T.T.T. and T.A.; visualization, D.M. and M.P.; supervision, D.M.; project administration, D.M., Z.T., T.T.T. and T.A.; funding acquisition, D.M., Z.T., T.T.T. and T.A. All authors have read and agreed to the published version of the manuscript.

Funding: This study was fully supported by Croatian Science Foundation (IP-2019–04–6183).

Institutional Review Board Statement: Not applicable.

Informed Consent Statement: Not applicable.

Data Availability Statement: The datasets generated during and/or analyzed during the current study are available from the corresponding author on reasonable request.

Acknowledgments: The authors gratefully acknowledge the donation of Cention and a curing unit by Ivoclar Vivadent.

Conflicts of Interest: The authors declare no conflict of interest. The funders had no role in the design of the study; in the collection, analyses, or interpretation of data; in the writing of the manuscript, or in the decision to publish the results.

References

1. Marovic, D.; Haugen, H.J.; Negovetic Mandic, V.; Par, M.; Zheng, K.; Tarle, Z.; Boccaccini, A.R. Incorporation of Copper-Doped Mesoporous Bioactive Glass Nanospheres in Experimental Dental Composites: Chemical and Mechanical Characterization. *Materials* **2021**, *14*, 2611. [CrossRef] [PubMed]
2. Marovic, D.; Tarle, Z.; Hiller, K.A.; Muller, R.; Ristic, M.; Rosentritt, M.; Skrtic, D.; Schmalz, G. Effect of silanized nanosilica addition on remineralizing and mechanical properties of experimental composite materials with amorphous calcium phosphate. *Clin. Oral Investig.* **2014**, *18*, 783–792. [CrossRef] [PubMed]
3. Marovic, D.; Tarle, Z.; Hiller, K.A.; Muller, R.; Rosentritt, M.; Skrtic, D.; Schmalz, G. Reinforcement of experimental composite materials based on amorphous calcium phosphate with inert fillers. *Dent. Mater.* **2014**, *30*, 1052–1060. [CrossRef]
4. Par, M.; Attin, T.; Tarle, Z.; Taubock, T.T. A New Customized Bioactive Glass Filler to Functionalize Resin Composites: Acid-Neutralizing Capability, Degree of Conversion, and Apatite Precipitation. *J. Clin. Med.* **2020**, *9*, 1173. [CrossRef]
5. Par, M.; Spanovic, N.; Mohn, D.; Attin, T.; Taubock, T.T.; Tarle, Z. Curing potential of experimental resin composites filled with bioactive glass: A comparison between Bis-EMA and UDMA based resin systems. *Dent. Mater.* **2020**, *36*, 711–723. [CrossRef] [PubMed]
6. Vallittu, P.K.; Boccaccini, A.R.; Hupa, L.; Watts, D.C. Bioactive dental materials-Do they exist and what does bioactivity mean? *Dent. Mater.* **2018**, *34*, 693–694. [CrossRef]
7. Darvell, B. Bioactivity—symphony or cacophony? A personal view of a tangled field. *Prosthesis* **2021**, *3*, 8. [CrossRef]

8. Hench, L.L. The story of Bioglass®. *J. Mater. Sci. Mater. Med.* **2006**, *17*, 967–978. [CrossRef]
9. Spagnuolo, G. Bioactive Dental Materials: The Current Status. *Materials* **2022**, *15*, 2016. [CrossRef]
10. Francois, P.; Fouquet, V.; Attal, J.-P.; Dursun, E. Commercially Available Fluoride-Releasing Restorative Materials: A Review and a Proposal for Classification. *Materials* **2020**, *13*, 2313. [CrossRef]
11. Watson, T.F.; Atmeh, A.R.; Sajini, S.; Cook, R.J.; Festy, F. Present and future of glass-ionomers and calcium-silicate cements as bioactive materials in dentistry: Biophotonics-based interfacial analyses in health and disease. *Dent. Mater.* **2014**, *30*, 50–61. [CrossRef]
12. Nicholson, J.W.; Wilson, A.D. The effect of storage in aqueous solutions on glass-ionomer and zinc polycarboxylate dental cements. *J. Mater. Sci. Mater. Med.* **2000**, *11*, 357–360. [CrossRef]
13. Gemalmaz, D.; Yoruc, B.; Ozcan, M.; Alkumru, H.N. Effect of early water contact on solubility of glass ionomer luting cements. *J. Prosthet. Dent.* **1998**, *80*, 474–478. [CrossRef]
14. Lohbauer, U. Dental Glass Ionomer Cements as Permanent Filling Materials?—Properties, Limitations Future Trends. *Materials* **2009**, *3*, 76–96. [CrossRef]
15. Xie, D.; Brantley, W.A.; Culbertson, B.M.; Wang, G. Mechanical properties and microstructures of glass-ionomer cements. *Dent. Mater.* **2000**, *16*, 129–138. [CrossRef]
16. Paolone, G. Direct composites in anteriors: A matter of substrate. *Int. J. Esthet. Dent.* **2017**, *12*, 468–481.
17. Marovic, D.; Taubock, T.T.; Attin, T.; Panduric, V.; Tarle, Z. Monomer conversion and shrinkage force kinetics of low-viscosity bulk-fill resin composites. *Acta Odontol. Scand.* **2015**, *73*, 474–480. [CrossRef]
18. Marovic, D.; Par, M.; Macan, M.; Klaric, N.; Plazonic, I.; Tarle, Z. Aging-Dependent Changes in Mechanical Properties of the New Generation of Bulk-Fill Composites. *Materials* **2022**, *15*, 902. [CrossRef]
19. Garoushi, S.; Vallittu, P.K.; Lassila, L. Characterization of fluoride releasing restorative dental materials. *Dent. Mater. J.* **2018**, *37*, 293–300. [CrossRef]
20. Eweis, A.H.; Yap, A.U.; Yahya, N.A. Impact of dietary solvents on flexural properties of bulk-fill composites. *Saudi Dent. J.* **2018**, *30*, 232–239. [CrossRef]
21. Ozer, F.; Irmak, O.; Yakymiv, O.; Mohammed, A.; Pande, R.; Saleh, N.; Blatz, M. Three-year Clinical Performance of Two Giomer Restorative Materials in Restorations. *Oper. Dent.* **2021**, *46*, E60–E67. [CrossRef] [PubMed]
22. Par, M.; Gubler, A.; Attin, T.; Tarle, Z.; Taubock, T.T. Ion release and hydroxyapatite precipitation of resin composites functionalized with two types of bioactive glass. *J. Dent.* **2022**, *118*, 103950. [CrossRef]
23. Tiskaya, M.; Al-Eesa, N.A.; Wong, F.S.L.; Hill, R.G. Characterization of the bioactivity of two commercial composites. *Dent. Mater.* **2019**, *35*, 1757–1768. [CrossRef] [PubMed]
24. Par, M.; Gubler, A.; Attin, T.; Tarle, Z.; Taubock, T.T. Anti-demineralizing protective effects on enamel identified in experimental and commercial restorative materials with functional fillers. *Sci. Rep.* **2021**, *11*, 11806. [CrossRef] [PubMed]
25. Par, M.; Gubler, A.; Attin, T.; Tarle, Z.; Tarle, A.; Taubock, T.T. Experimental Bioactive Glass-Containing Composites and Commercial Restorative Materials: Anti-Demineralizing Protection of Dentin. *Biomedicines* **2021**, *9*, 1616. [CrossRef] [PubMed]
26. Tecco, S.; Grusovin, M.G.; Sciara, S.; Bova, F.; Pantaleo, G.; Cappare, P. The association between three attitude-related indexes of oral hygiene and secondary implant failures: A retrospective longitudinal study. *Int. J. Dent. Hyg.* **2018**, *16*, 372–379. [CrossRef]
27. Gherlone, E.; Polizzi, E.; Tete, G.; Cappare, P. Dentistry and Covid-19 pandemic: Operative indications post-lockdown. *New Microbiol.* **2021**, *44*, 1–11.
28. Akashi, A.; Matsuya, Y.; Unemori, M.; Akamine, A. The relationship between water absorption characteristics and the mechanical strength of resin-modified glass-ionomer cements in long-term water storage. *Biomaterials* **1999**, *20*, 1573–1578. [CrossRef]
29. Marovic, D.; Par, M.; Crnadak, A.; Sekelja, A.; Negovetic Mandic, V.; Gamulin, O.; Rakic, M.; Tarle, Z. Rapid 3 s Curing: What Happens in Deep Layers of New Bulk-Fill Composites? *Materials* **2021**, *14*, 515. [CrossRef]
30. Haugen, H.J.; Marovic, D.; Par, M.; Thieu, M.K.L.; Reseland, J.E.; Johnsen, G.F. Bulk Fill Composites Have Similar Performance to Conventional Dental Composites. *Int. J. Mol. Sci.* **2020**, *21*, 5136. [CrossRef]
31. Bahari, M.; Kahnamoui, M.A.; Chaharom, M.E.E.; Kimyai, S.; Sattari, Z. Effect of curing method and thermocycling on flexural strength and microhardness of a new composite resin with alkaline filler. *Dent. Res. J.* **2021**, *18*, 96. [CrossRef]
32. Yap, A.U.; Choo, H.S.; Choo, H.Y.; Yahya, N.A. Flexural Properties of Bioactive Restoratives in Cariogenic Environments. *Oper. Dent.* **2021**, *46*, 448–456. [CrossRef] [PubMed]
33. Ilie, N. Comparative Effect of Self- or Dual-Curing on Polymerization Kinetics and Mechanical Properties in a Novel, Dental-Resin-Based Composite with Alkaline Filler. *Materials* **2018**, *11*, 108. [CrossRef]
34. Gomes de Araújo-Neto, V.; Sebold, M.; Fernandes de Castro, E.; Feitosa, V.P.; Giannini, M. Evaluation of physico-mechanical properties and filler particles characterization of conventional, bulk-fill, and bioactive resin-based composites. *J. Mech. Behav. Biomed. Mater.* **2021**, *115*, 104288. [CrossRef]
35. Kelić, K.; Par, M.; Peroš, K.; Šutej, I.; Tarle, Z. Fluoride-Releasing Restorative Materials: The Effect of a Resinous Coat on Ion Release. *Acta Stomatol. Croat.* **2020**, *54*, 371–381. [CrossRef] [PubMed]
36. Roulet, J.F.; Gummadi, S.; Hussein, H.S.; Abdulhameed, N.; Shen, C. In vitro wear of dual-cured bulkfill composites and flowable bulkfill composites. *J. Esthet. Restor. Dent.* **2020**, *32*, 512–520. [CrossRef]
37. Pribadi, N.; Budiarti, D.; Kurniawan, H.J.; Widjiastuti, I. The NF-kB and Collagen Type 1 Expression in Dental Pulp after Treated Calcium Hydroxide Combined with Propolis. *Eur. J. Dent.* **2021**, *15*, 122–126. [CrossRef]

38. Aldhafyan, M.; Silikas, N.; Watts, D.C. Influence of curing modes on conversion and shrinkage of dual-cure resin-cements. *Dent. Mater.* **2022**, *38*, 194–203. [CrossRef]
39. *ISO 4049:2009*; Dentistry—Polymer-Based Filling, Restorative and Luting Materials. International Organization for Standardization: Geneva, Switzerland, 2000.
40. Stansbury, J.W. Dimethacrylate network formation and polymer property evolution as determined by the selection of monomers and curing conditions. *Dent. Mater.* **2012**, *28*, 13–22. [CrossRef] [PubMed]
41. Barszczewska-Rybarek, I.M. A Guide through the Dental Dimethacrylate Polymer Network Structural Characterization and Interpretation of Physico-Mechanical Properties. *Materials* **2019**, *12*, 4057. [CrossRef]
42. Sirovica, S.; Solheim, J.H.; Skoda, M.W.A.; Hirschmugl, C.J.; Mattson, E.C.; Aboualizadeh, E.; Guo, Y.; Chen, X.; Kohler, A.; Romanyk, D.L.; et al. Origin of micro-scale heterogeneity in polymerisation of photo-activated resin composites. *Nat. Commun.* **2020**, *11*, 1849. [CrossRef]
43. Da Silva, E.M.; Poskus, L.T.; Guimaraes, J.G.; de Araujo Lima Barcellos, A.; Fellows, C.E. Influence of light polymerization modes on degree of conversion and crosslink density of dental composites. *J. Mater. Sci. Mater. Med.* **2008**, *19*, 1027–1032. [CrossRef]
44. Soh, M.S.; Yap, A.U. Influence of curing modes on crosslink density in polymer structures. *J. Dent.* **2004**, *32*, 321–326. [CrossRef] [PubMed]
45. Sideridou, I.D.; Karabela, M.M.; Bikiaris, D.N. Aging studies of light cured dimethacrylate-based dental resins and a resin composite in water or ethanol/water. *Dent. Mater.* **2007**, *23*, 1142–1149. [CrossRef] [PubMed]
46. Yap, A.U.; Wang, X.; Wu, X.; Chung, S.M. Comparative hardness and modulus of tooth-colored restoratives: A depth-sensing microindentation study. *Biomaterials* **2004**, *25*, 2179–2185. [CrossRef]
47. Bhatia, H.P.; Singh, S.; Sood, S.; Sharma, N. A Comparative Evaluation of Sorption, Solubility, and Compressive Strength of Three Different Glass Ionomer Cements in Artificial Saliva: An in vitro Study. *Int. J. Clin. Pediatr. Dent.* **2017**, *10*, 49–54. [CrossRef] [PubMed]
48. Mustafa, R.; Alshali, R.Z.; Silikas, N. The effect of desiccation on water sorption, solubility and hygroscopic volumetric expansion of dentine replacement materials. *Dent. Mater.* **2018**, *34*, e205–e213. [CrossRef]
49. Cefaly, D.F.; Franco, E.B.; Mondelli, R.F.; Francisconi, P.A.; Navarro, M.F. Diametral tensile strength and water sorption of glass-ionomer cements used in Atraumatic Restorative Treatment. *J. Appl. Oral Sci.* **2003**, *11*, 96–101. [CrossRef]
50. Muller, J.A.; Rohr, N.; Fischer, J. Evaluation of ISO 4049: Water sorption and water solubility of resin cements. *Eur. J. Oral Sci.* **2017**, *125*, 141–150. [CrossRef] [PubMed]
51. Par, M.; Spanovic, N.; Bjelovucic, R.; Marovic, D.; Schmalz, G.; Gamulin, O.; Tarle, Z. Long-term water sorption and solubility of experimental bioactive composites based on amorphous calcium phosphate and bioactive glass. *Dent. Mater. J.* **2019**, *38*, 555–564. [CrossRef]
52. Alshali, R.Z.; Salim, N.A.; Satterthwaite, J.D.; Silikas, N. Long-term sorption and solubility of bulk-fill and conventional resin-composites in water and artificial saliva. *J. Dent.* **2015**, *43*, 1511–1518. [CrossRef] [PubMed]
53. Ferracane, J.L.; Berge, H.X.; Condon, J.R. In vitro aging of dental composites in water—effect of degree of conversion, filler volume, and filler/matrix coupling. *J. Biomed. Mater. Res.* **1998**, *42*, 465–472. [CrossRef]
54. Par, M.; Tarle, Z.; Hickel, R.; Ilie, N. Mechanical properties of experimental composites containing bioactive glass after artificial aging in water and ethanol. *Clin. Oral Investig.* **2019**, *23*, 2733–2741. [CrossRef] [PubMed]
55. Szczesio-Wlodarczyk, A.; Sokolowski, J.; Kleczewska, J.; Bociong, K. Ageing of Dental Composites Based on Methacrylate Resins—A Critical Review of the Causes and Method of Assessment. *Polymers* **2020**, *12*, 882. [CrossRef] [PubMed]
56. Heintze, S.D.; Ilie, N.; Hickel, R.; Reis, A.; Loguercio, A.; Rousson, V. Laboratory mechanical parameters of composite resins and their relation to fractures and wear in clinical trials—A systematic review. *Dent. Mater.* **2017**, *33*, e101–e114. [CrossRef]

Article

Short and Long-Term Solubility, Alkalizing Effect, and Thermal Persistence of Premixed Calcium Silicate-Based Sealers: AH Plus Bioceramic Sealer vs. Total Fill BC Sealer

David Donnermeyer [1,*], Patrick Schemkämper [1], Sebastian Bürklein [2] and Edgar Schäfer [2]

[1] Department of Periodontology and Operative Dentistry, Westphalian Wilhelms-University, Albert-Schweitzer-Campus 1, Building W 30, 48149 Münster, Germany
[2] Central Interdisciplinary Ambulance in the School of Dentistry, Albert-Schweitzer-Campus 1, Building W 30, 48149 Münster, Germany
* Correspondence: david.donnermeyer@ukmuenster.de; Tel.: +49-251-8347064

Abstract: This study aimed to investigate the short- and long-term solubility, alkalizing potential, and suitability for warm-vertical compaction of AH Plus Bioceramic Sealer (AHBC), and Total Fill BC Sealer (TFBC) compared to the epoxy-resin sealer AH Plus (AHP). A solubility test was performed according to ISO specification 6876 and further similar to ISO requirements over a period of 1 month in distilled water (AD) and 4 months in phosphate-buffered saline (PBS). The pH of the immersion solution was determined weekly. Sealers were exposed to thermal treatment at 37, 57, 67, and 97 °C for 30 s. Furthermore, heat treatment at 97 °C was performed for 180 s to simulate inappropriate implementation of warm vertical filling techniques. Physical properties (setting time, flow, film thickness according to ISO 6876) and chemical properties (Fourier transformed infrared spectroscopy) were assessed. AHBC and TFBC were associated with significantly higher solubility than AHP over 1 month in AD and 4 months in PBS ($p < 0.05$). AHBC and TFBC both reached high initial alkaline pH, while TFBC was associated with a longer-lasting alkaline potential than AHBC. Initial pH of AHBC and TFBC was higher in AD than in PBS. None of the sealers were compromised by thermal treatment.

Keywords: AH plus Bioceramic Sealer; alkalizing potential; pH; solubility; Total Fill BC Sealer; warm vertical compaction

Citation: Donnermeyer, D.; Schemkämper, P.; Bürklein, S.; Schäfer, E. Short and Long-Term Solubility, Alkalizing Effect, and Thermal Persistence of Premixed Calcium Silicate-Based Sealers: AH Plus Bioceramic Sealer vs. Total Fill BC Sealer. *Materials* **2022**, *15*, 7320. https://doi.org/10.3390/ma15207320

Academic Editors: Tobias Tauböck and Matej Par

Received: 28 September 2022
Accepted: 14 October 2022
Published: 19 October 2022

Publisher's Note: MDPI stays neutral with regard to jurisdictional claims in published maps and institutional affiliations.

Copyright: © 2022 by the authors. Licensee MDPI, Basel, Switzerland. This article is an open access article distributed under the terms and conditions of the Creative Commons Attribution (CC BY) license (https://creativecommons.org/licenses/by/4.0/).

1. Introduction

Calcium silicate-based sealers have emerged as a relevant alternative to epoxy resin sealers in the past decade. Clinical studies have reported on the successful implementation of premixed calcium silicate-based sealers in root canal obturation [1,2]. Due to their beneficial properties concerning antimicrobial activity, biocompatibility, and bioactivity, these sealers have changed the perspective on root canal obturation, but have also demanded new concepts because their effects mainly rely on a high proportion of sealer inside the root canal filling [3].

Most of their beneficial properties are based on the solubility of reactional by-products of calcium silicates over a period of several weeks [4]. Mainly, the dissolution of calcium hydroxide during the initial setting reaction of calcium silicates with water is the principle of the advantageous properties [5]. While there is consensus that the aim of a root canal obturation should be a long-lasting fluid and bacteria tight seal of the root canal system, drawbacks concerning solubility of calcium silicate-based sealers are a matter of discussion [6]. High solubility could result in a weaker seal of the root canal system, allowing tissue fluid to leak into the apical region of the root canal system and byproducts of trapped bacteria to leak out of the root canal system [6]. A smaller proportion of sealer achieved by warm compaction of the gutta-percha core materials could address this problem. While this would adversely compromise the beneficial properties of calcium-silicate based sealers, it is also necessary to investigate the thermal stability of sealers before subjecting them to such techniques [7]. Destruction of the sealer component's chemical structure could result in insufficient root canal obturation due to incomplete setting. In addition, the changes of physical properties, e.g., flow or film thickness, would lead to insufficient root canal obturation because the sealer may not be able to reach the complete complex anatomy of the root canal system.

Recently, a new premixed calcium silicate-based sealer, AH Plus Bioceramic (AHBC, Dentsply Sirona, York, PA, USA), was introduced. While it contains only tricalcium silicate as a reactive component and not di- and tri-calcium silicates like most other calcium silicate-based materials such as Total Fill BC Sealer (TFBC; FKG Dentaire, La Chaux-des-Fonds, Switzerland), AHBC comprises dimethyl sulfoxide as a filler, which is not known from other calcium silicate-based sealers. This results in a lower proportion of calcium silicates than in other premixed sealers like TFBC. No data exist to date addressing the formulation of AHBC in terms of its solubility and alkalizing potential over short and long periods and its suitability for warm obturation techniques.

The aim of this study was to measure the short- and long-term solubility, pH, and thermal stability of the new AHBC compared to a contemporary well-investigated calcium-silicate-based sealer TFBC and the epoxy resin-based sealer AH Plus (AHP, Dentsply Sirona).

2. Materials and Methods

AH Plus Bioceramic Sealer and Total Fill BC Sealer were investigated. Both sealers are premixed products, and no preparations were needed. AH Plus, which was mixed using the AH Plus Jet, served as the control.

2.1. Sample Size Calculation

Power calculation using G*Power 3.1 (Heinrich Heine University, Düsseldorf, Germany) indicated a sample size of at least nine samples per group. Thus, 10 samples were prepared per group for solubility evaluation. Concerning the physical properties after thermal treatment, three tests were carried out for each temperature level and each sealer and the mean was calculated according to ISO 6876 [8].

2.2. Solubility (Long-Term)

To evaluate the long-term solubility, sealer specimens were immersed in distilled water (AD) and in phosphate buffered saline solution (PBS, Pharmacy of the University Hospital, Münster, Germany), and the specimens' change in weight was recorded in a modification of a methodology described previously [4]. Stainless steel ring washers (height 1.6 ± 0.1 mm, internal diameter 20.0 ± 0.1 mm) were cleaned in an ultrasound bath with acetone for 15 min and a teflon band was fixed on each washer. The prepared washers were weighed three times (accuracy ± 0.0001 g; Sartorius 1801 MPS, Göttingen, Germany), and the mean was calculated. The washers were placed on a glass plate and filled to slight excess with sealer dispensed from the syringes. To ensure complete setting of all sealers before testing, glass plates and samples were placed on a gauze immersed in physiological solution (PBS) in a closed container at 37 °C for 24 h. The proper setting was evaluated in preliminary experiments. After setting of the sealers, excess material was trimmed to the surface level of the washer by using silicon carbide paper (600 grit). The specimens were weighed three times before the immersion of the samples and the sealer weight was calculated. Twenty samples were prepared for immersion in AD and 40 samples for immersion in PBS (150 mL) were prepared from each sealer. Each of the 10 samples were immersed in AD for 14 and 28 days and in PBS for 24 h, 14 and 28 days, and 2 and 4 months. Twenty washers for each group were prepared for immersion in AD or PBS (n = 10) without any sealer as the negative control during the entire period of 1 and 4 months, respectively. All samples were stored in an incubator (Heraeus, Hanau, Germany) at 37 °C and 100% relative humidity. After 24 h, the first fluid change was performed on all samples and all fluids were changed every 7 days thereafter. After immersion, the samples were weighed again three times, and the mass of the sealers was determined. The difference between the original weight of the material and its final weight was recorded and the percentual mass loss was calculated as solubility.

2.3. Solubility (Short-Term)

A solubility test was carried out over 24 h according to ISO specification 6876 in AD and in PBS. Sealer specimens were prepared in ring molds as stated by ISO 6876 specification. After determination of the sealer mass (accuracy ± 0.0001 g; Sartorius 1801MPS), 2 specimens of each sealer were immersed in 50 mL AD in a covered dish and placed in an incubator (Heraeus) at 37°C and 100% humidity. After 24 h, the specimens were washed with AD and dried. The samples were weighed 3 times, and the mean mass of the sealers was determined. The difference between the original weight of the material and its final weight was recorded and the percentual mass loss was calculated as solubility.

2.4. pH

The pH value assessment was performed parallel to the solubility test [4]. The pH value was determined with an electrode pH meter (PB 11, Sartorius, Göttingen, Germany). The accuracy of the pH meter was controlled with calibration solutions (pH 4, 7, and 10; Merck, Darmstadt, Germany). After each individual measurement, the electrode was flushed with AD. The pH measurement was carried out after 24 h, and weekly before renewal of the test liquids at 37 °C fluid temperature.

2.5. Thermal Treatment—Physical Properties

Setting time, film thickness, and flow were assessed similar to ISO specification 6876 and after thermal treatment, as described previously [7,9]. Portions of 0.5 mL of each sealer were dispensed directly into a 2 mL plastic tube (Safe-Lock Tubes, Eppendorf, Hamburg, Germany). A K-type thermocouple (GHM Messtechnik, Regenstauf, Germany) was placed into the sealer, and the samples were heated in a thermo-controlled water bath until temperatures of 37 °C, 57 °C, 67 °C, and 97 °C were achieved inside the samples. These temperatures were selected in accordance with recently published data [7,9,10]. The temperature of the sealer was controlled by GSVmulti software (version 1.27, ME-Meßsysteme, Hennigsdorf, Germany) at a frequency of 50 Hz using the thermocouple. All samples were retained for 30 s at the respective temperatures and were cooled to 37 °C in a second water bath afterward. For the evaluation of the influence of elongated heating, sealers were also heated to 97 °C for 180 s. The described procedure took about 3 min.

The setting time was assessed by dispensing the preheated sealer specimens onto glass plates inside a stainless-steel ring (d = 10 mm, h = 2 mm). After transfer to an incubator at 37 °C and 100% humidity, a stopwatch was used to determine the setting time of the material. A cylindrical indenter with a flat end tip diameter of 2 mm and a mass of 100 g was used as defined in ISO 6876. The materials setting point was defined as the point when the needle left no indentation on the sealer's surface anymore. A film thickness test was carried out similar to ISO 6876 with slight modifications of the temporal process due to the preheating of the sealers. After the thermal treatment, a portion of each specimen was placed on a glass plate measuring 40 mm × 40 mm and 5 mm in thickness. A second glass plate of 5 mm thickness and a surface area of 200 mm^2 was placed centrally on top. A load of 150 N was generated vertically on the top plate by a universal testing machine (Lloyd LF Plus, Ametek, Berwyn, PA, USA) for 10 min. The thickness of the two assembled glass plates was measured before each test and after the testing procedure using a digital micrometer. Due to a higher viscosity of the sealers reported at high temperatures [9], the sealers were portioned by weight instead of volume. Using a precision scale and a graduated pipette, 0.05 mL of sealer was found to correspond 0.1285 g of AHBC, 0.1265 g of TFBC [9], and 0.140 g of AHP [7], respectively, at 20 °C. A portion of each specimen was placed on a glass plate measuring 40 mm × 40 mm and 5 mm in thickness. A second glass plate with the same dimension and a weight resulting in a total mass of 120 g were placed on top centrally and the assembly was left for 10 min. The maximum and minimum diameters of the compressed sealer phase were measured using a digital caliper. If the maximum and minimum diameters were within 1 mm, the mean was calculated. Three tests were carried out for each temperature level.

2.6. Thermal Treatment—Chemical Properties

For Fourier transform infrared spectroscopy, the specimens were stored on glass plates in an incubator for 8 weeks at 37 °C and 100% humidity. The set specimens were powdered using a mortar. Then, 0.002 g of sealer powder were added to 0.2 g potassium bromide and pressed to a pill. Fourier transform infrared spectroscopy was performed using the Vertex 70v with a mercury cadmium telluride MCT detector (Bruker, Billerica, MA, USA) by 256 scans per test 2 times at each temperature level. One result was selected for evaluation in case no difference occurred between the spectra [9].

2.7. Statistical Analysis

Data of solubility were normally distributed (Kolmogorov–Smirnov-test) and analyzed with ANOVA and Scheffé post hoc test ($p = 0.05$). Data concerning physical properties (setting time, film thickness, and flow) were analyzed using Kruskal–Wallis test at $p = 0.05$.

3. Results

3.1. Solubility (Long-Term)

After 14 days and 28 days AHBC and TFBC showed higher solubility (about 30%) in AD with an increase over time, while AHP was not associated with relevant solubility. The difference between the calcium silicate-based sealers AHBC and TFBC was significant at 14 and 28 days ($p < 0.05$). After 28 days, TFBC was associated with significantly higher solubility than AHBC ($p < 0.05$), while no such difference was observed after 14 days. Immersed in PBS, the solubility of AHBC and TFBC was lower at 14 and 28 days compared to AD. Over a 4-month period in PBS, the solubility of AHBC and TFBC was significantly higher than of AHP at all measurement times ($p < 0.05$). Significant differences between AHBC and TFBC were only detected after 14 days in PBS, when AHBC presented with significantly higher solubility ($p < 0.05$) (Table 1).

Table 1. Means and standard deviations: solubility of AH Plus Bioceramic Sealer, Total Fill BC Sealer, and AH Plus in AD over 28 days and in PBS over 4 months, respectively. Superscript letters indicate statistically significant differences at measurement dates in AD and PBS ($p < 0.05$).

	AD			PBS		
	AH Plus Bioceramic Sealer	Total Fill BC Sealer	AH Plus	AH Plus Bioceramic Sealer	Total Fill BC Sealer	AH Plus
14 days	30.44 ± 1.00 [A]	32.75 ± 5.26 [A]	0.55 ± 0.17 [B]	19.24 ± 2.56 [A]	14.05 ± 2.35 [B]	0.02 ± 0.23 [C]
28 days	33.09 ± 0.81 [B]	35.55 ± 1.35 [A]	0.48 ± 0.20 [C]	20.80 ± 2.01 [A]	20.64 ± 2.87 [A]	0.28 ± 0.16 [B]
2 months				16.82 ± 2.38 [A]	14.78 ± 4.02 [A]	0.30 ± 0.18 [B]
4 months				18.40 ± 1.91 [A]	20.50 ± 9.23 [A]	0.32 ± 0.08 [B]

3.2. Solubility (Short-Term)

The results of the solubility test according to ISO 6876 are presented in Table 2. While AH Plus presented with negligible weight loss both in AD and PBS, the calcium silicate-based sealer AHBC and TFBC were associated with relevant loss up to 34.3%. The solubility of AHBC and TBC presented similarly high in AD and PBS after 24 h.

Table 2. Solubility of AH Plus Bioceramic Sealer, Total Fill BC Sealer, and AH Plus after 24 h in AD according to ISO 6876, and in PBS.

	AH Plus Bioceramic Sealer		Total Fill BC Sealer		AH Plus	
	AD	PBS	AD	PBS	AD	PBS
Solubility (%)	33.2	33.7	28.7	32.1	0.4	0.5

3.3. pH

AHBC and TFBC reached high pH values above 12 after 24 h in AD. The pH in AD constantly decreased over 1 month with AHBC showing a more pronounced decrease. Immersed in PBS, both sealers reached high initial pH values. The pH of TFBC decreased constantly over a period of 3 months until no relevant alkalization of the buffer solution was measured. A faster pH decrease was observed, with AHBC reaching close to the baseline pH after 1.5 months already (Figure 1). AHP did not influence the pH of the immersion solutions.

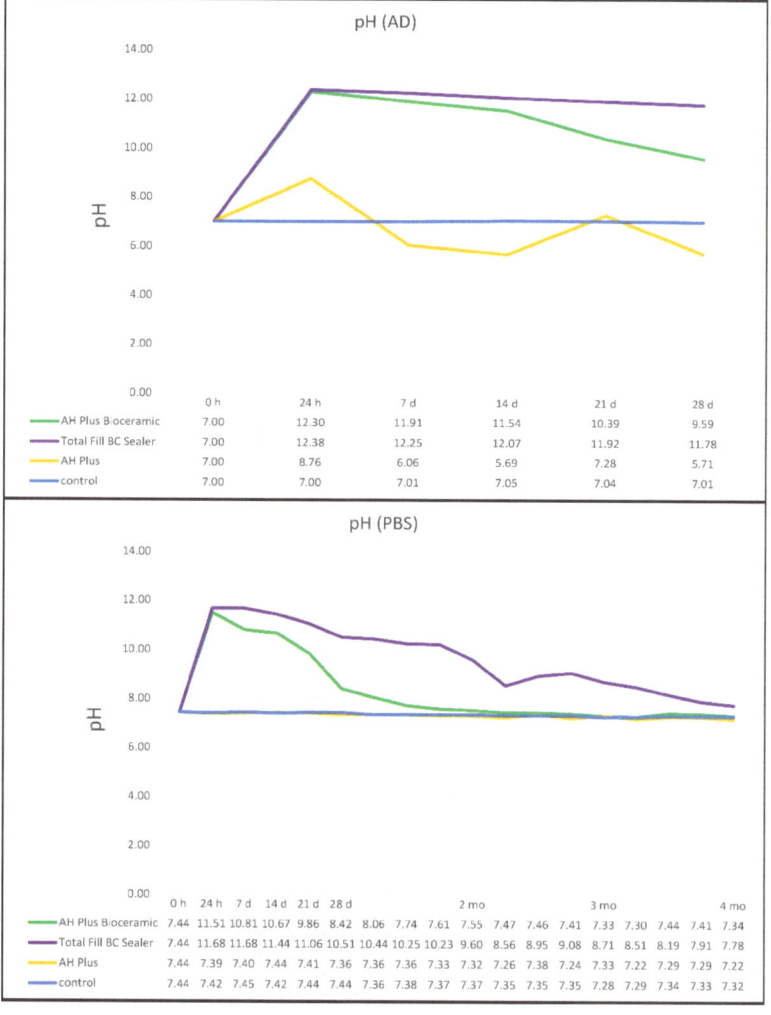

Figure 1. pH of AH Plus Bioceramic Sealer, Total Fill BC Sealer, and AH Plus in distilled water (AD) over 28 days and in PBS over 4 months, respectively.

3.4. Thermal Treatment—Physical Properties

The setting time, film thickness, and flow of all sealers were not relevantly influenced by any thermal treatment and did not exceed clinically relevant and ISO-defined thresholds [8] (Tables 3–5). Significant differences in the setting time were observed for all sealers ($p < 0.05$), but none of them were following a pattern. The film thickness of neither AHBC, TFBC, nor AHP was affected significantly by thermal treatment ($p > 0.05$). The flow of AHBC decreased with increasing temperature. Significant differences only occurred between 37 °C (30 s) and 97 °C (180 s) ($p < 0.05$). The flow of TFBC slightly decreased with thermal exposure, showing statistically significant differences between 57 °C (30 s) and 97 °C (180 s) ($p < 0.05$). No statistically significant changes of flow were observed for AHP ($p > 0.05$).

Table 3. Physical properties in accordance with ISO 6876 of AH Plus Bioceramic (means and standard deviations (SD)) after thermal treatment. Statistical analysis of setting time, film thickness, and flow for AH Plus Bioceramic was performed by Kruskal–Wallis test ($p < 0.05$).

	Group Number	Setting Time (h)			Film Thickness (m)			Flow (mm)		
		Mean	SD	Different from Group Number	Mean	SD	Different from Group Number	Mean	SD	Different from Group Number
37 (30 s)	1	9.861	0.369		0.016	0.007		25.7	1.0	5
57 (30 s)	2	10.472	0.243		0.020	0.011		25.5	2.1	
67 (30 s)	3	11.156	0.184	5	0.022	0.008		22.8	0.7	
97 (30 s)	4	10.850	0.200		0.015	0.002		18.6	0.5	
97 (180 s)	5	9.200	0.225	3	0.017	0.005		16.1	0.4	1

Table 4. Physical properties in accordance with ISO 6876 of Total Fill BC Sealer (means and standard deviations (SD)) after thermal treatment. Statistical analysis of setting time, film thickness, and flow for Total Fill BC Sealer was performed by Kruskal–Wallis test ($p < 0.05$).

	Group Number	Setting Time (h)			Film Thickness (m)			Flow (mm)		
		Mean	SD	Different from Group Number	Mean	SD	Different from Group Number	Mean	SD	Different from Group Number
37 (30 s)	1	24.383	0.166	5	0.018	0.002		25.1	0.7	
57 (30 s)	2	23.850	0.350		0.020	0.002		26.3	1.4	5
67 (30 s)	3	23.507	0.081		0.019	0.003		25.0	0.6	
97 (30 s)	4	22.897	0.387		0.017	0.001		23.2	0.3	
97 (180 s)	5	21.303	0.160	1	0.017	0.002		21.0	0.7	2

Table 5. Physical properties in accordance with ISO 6876 of AH Plus (means and standard deviations (SD)) after thermal treatment. Statistical analysis of setting time, film thickness, and flow for AH Plus was performed by Kruskal–Wallis test ($p < 0.05$).

	Group Number	Setting Time (h)			Film Thickness (m)			Flow (mm)		
		Mean	SD	Different from Group Number	Mean	SD	Different from Group Number	Mean	SD	Different from Group Number
37 (30 s)	1	9.77	0.30	5	0.027	0.002		22.1	0.5	
57 (30 s)	2	9.59	0.15		0.028	0.007		23.3	0.4	
67 (30 s)	3	8.61	0.21		0.026	0.003		23.1	0.8	
97 (30 s)	4	8.14	0.25		0.025	0.001		24.8	0.7	
97 (180 s)	5	7.41	0.28	1	0.026	0.001		23.2	0.9	

3.5. Thermal Treatment—Chemical Properties

No changes of the chemical structure of AHBC, TFBC, and AHP were indicated by the spectroscopic plots of FT-IR spectroscopy at any thermal treatment level (Figure 2). Both AHBC and TFBC spectroscopic plots indicated the presence of water by a broad absorption band around 3400 cm^{-1} and a peak at 1650 cm^{-1} [9,11]. Carbonates were detected for AHBC and TFBC at 878 cm^{-1} and between 1400 and 1500 cm^{-1} [12]. A calcium hydroxide band (O-H-stretch at 3646 cm^{-1}) was not detected in AHBC and TFBC. Absorption between 970 and 1000 cm^{-1} was observed with AHBC and TFBC, as this indicated the formation of calcium silicate hydrate [12]. Characteristic peaks at ~2874 cm^{-1} and 2923 cm^{-1}, which are assigned to symmetric stretching of -CH3 and C-H-stretching of -CH2-, respectively, were found in TFBC but not in AHBC [13]. A peak at 1044 cm^{-1} indicated the presence of dimethyl sulfoxide solely in AHBC specimens [14] (Figure 2).

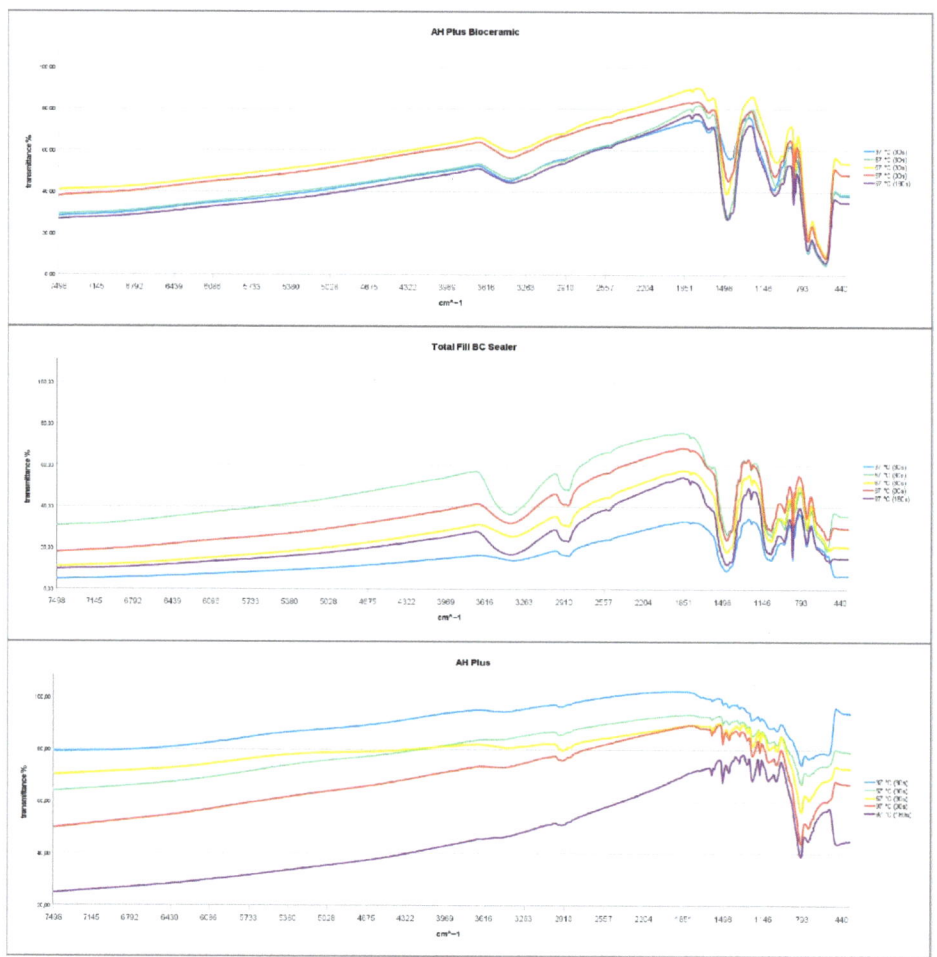

Figure 2. Spectroscopic plots of FT-IR spectroscopy after thermal treatment of AHBC, TFBC, and AHP.

4. Discussion

In previous studies evidence was found that the solubility, alkalizing potential, and bond strength of calcium silicate-based sealers depend on the immersion solution [4,15]. It was assumed that the precipitation of hydroxy apatite on the surface of calcium silicate-based materials after contact to phosphate concludes in a decrease of solubility [4]. A long-term study on the solubility and pH of the two-component calcium silicate-based sealer BioRoot RCS (Septodont, St. Maur-des-Fossés, France) corroborated this thesis. So, far no comparisons of long-term-solubility and pH investigations exist on premixed calcium silicate-based sealers. Therefore, the purpose of this study was to investigate the long-term solubility of premixed calcium-silicate-based sealer TFBC and AHBC in AD and PBS over 1 and 4 months, respectively, and to compare these results with a 24 h-testing protocol according to ISO specification 6876 in AD and PBS.

The solubility tests were performed similar to the ISO 6876 testing protocol. While the ISO specification 6876 demands solubility testing in AD only, the test was also performed in PBS. As immersion in AD is not capable of predicting solubility in isotonic body fluids in the in vivo situation [16], PBS was used to simulate a proper environment. In addition, the ISO specification 6876 demands testing only for a period of 24 h, while calcium silicate-based materials are known for prolonged setting reactions [17]. Therefore, tests were performed in AD for up to 28 days and in PBS for 4 months to evaluate the clinical effect of solubility and alkalization.

In general, solubility is investigated in AD after 24 h of immersion. Still, the results published on solubility of TFBC show a wide range [18]. While some studies reported solubility less than 3% [19,20], others found solubility of more than 20% [21]. According to the present results, the solubility of TFBC was high both in AD and PBS with 28.7% and 32.1%, respectively. Subsequently, AHBC was associated with high solubility of 33.2% (AD) and 33.7% (PBS) after 24 h. The type of immersion solution did not influence the high initial solubility of TFBC and AHBC. In the follow-up period of 28 days in AD, the solubility of AHBC and TFBC did not relevantly increase, indicating that premixed calcium silicate-based sealers show high solubility during the initial setting phase and are stable hereafter. Still, the solubility of TFBC was significantly higher than that of AHBC after 28 days in AD. A possible explanation could be the lower proportion of calcium silicates in AHBC. Corroborating results can be found after immersion in PBS, when the solubility did not increase over a 4-month period. Percentual solubility presented even lower after 14 days to 4 months compared to 24 h, which could be explained by the precipitation of hydroxy apatite on the specimens' surface increasing the sealers' weight.

AHBC and TFBC showed a high initial alkalizing potential after 24 h. In accordance with previous results, pH values determined in AD were higher than in PBS [4]. It was hypothesized that leaked calcium hydroxide from the sealer matrix, which is the main reason for the alkalizing potential of calcium silicate-based sealers, is buffered in PBS when it reacts with phosphate from the solution forming hydroxy apatite [5]. This reaction may not occur in AD, allowing more calcium hydroxide to dilute into the immersion solution.

In AD, TFBC was capable of remaining highly alkaline over the period of 28 days, while pH of AHBC started to decline after 14 days. This was also observed in PBS. After 6 weeks, AHBC nearly decreased to the initial pH of the PBS solution. In contrast, the pH of TFBC decreased more slowly, keeping the alkaline pH for nearly 4 months. Long-term alkalization could coincide with the formulation of the investigated sealers. While TFBC contains about 27 to 50% calcium silicates and 1 to 4% calcium hydroxide as stated by the manufacturer, the percentual proportion is only 5 to 15% calcium silicates in AHBC. The more calcium silicates are present, the more calcium hydroxide can be generated from their setting reaction. As the setting reaction of calcium silicates is known to last for several weeks, a higher proportion of calcium silicates could be an indicator for the longer lasting alkalization. A high pH is even prolonged when di-calcium silicates are present, as in TFBC, because di-calcium silicates present with slower reaction kinetics. Meanwhile, tri-calcium silicates, which are the only source of calcium hydroxide in AHBC, are more reactive in the initial phase of the calcium silicate setting reaction.

The alkaline pH caused by calcium silicate-based sealers is regarded as one of their major advantages. Calcium hydroxide is the major factor in biocompatibility as it leads to the formation of hydroxy apatite on the sealer surface after coming in contact with body fluid, and it also plays a role in the eradication of microorganisms still present after chemo-mechanical preparation in niches of the infected root canal system [22]. Calcium hydroxide also slowly affects microorganisms. Therefore, a long-lasting replenishment of calcium hydroxide from a calcium silicate-based sealer could be regarded beneficial and compensate for disadvantages such as high initial solubility. Furthermore, the alkaline pH is capable of inducing apical healing and mineralization of the apical alveolar bone structure. Accordingly, good biocompatibility was reported for AHBC and TFBC [23]. Still, a higher mineralization potential was associated with TFBC compared to AHBC [23], which is consistent with the higher and prolonged alkaline pH caused by the elution of calcium hydroxide observed in the present study.

When it comes to the investigation of the effect of heat on sealer stability, choosing a clinically relevant temperature and exposure period is crucial for interpretation of the results [10]. Thus, a range of temperatures and application times was investigated as described previously [7,9]. While a resistance to the thermal treatment of TFBC and AHP in terms of its physical properties and its chemical structure has been previously reported [7,9,11,24], no such data is available for the new AHBC. In addition to the stability of the physical properties and the FTIR spectra at all temperature levels, the presence of carbonates in the FTIR spectra is a sign of the formation of calcium hydroxide and indicates that the setting reaction of calcium silicates was not influenced by the thermal treatment. Calcium hydroxide reacts with carbon dioxide under atmospheric storage forming carbonates. Additionally, the bands indicating organic molecules are present in all spectra of AHBC and TFBC, indicating that the organic fillers used as thickening agents are able to withstand short period thermal stress. In accordance with the results for other premixed calcium silicate-based sealers [9,11,24,25], AHBC was found to be resistant against the thermal treatment performed in this study.

Premixed calcium silicate-based sealers presented with higher solubility and pH than the epoxy-resin sealer AH Plus. Still, among the premixed calcium silicate-based sealers, differences in the alkalizing potential both in AD and PBS were found to correspond to the sealer formulations. Premixed calcium silicate-based sealers were resistant to the heat ranges that occur during warm obturation techniques.

5. Conclusions

High solubility is inherent with premixed calcium silicate-based sealers. This results in high alkalizing potential, which is a major benefit in the application of calcium-silicate-based sealers. The higher the proportion of di- and tri-calcium silicates, the longer the alkaline pH can be observed. AHBC and TFBC can be considered as safe for warm-vertical compaction.

Author Contributions: Conceptualization, D.D. and E.S.; methodology, D.D.; software, E.S. and D.D.; validation, P.S. and E.S.; formal analysis, D.D.; investigation, P.S. and D.D.; resources, E.S.; data curation, D.D. and E.S.; writing—original draft preparation, D.D.; writing—review and editing, E.S. and S.B.; visualization, D.D.; supervision, D.D.; project administration, D.D. All authors have read and agreed to the published version of the manuscript.

Funding: This research received no external funding.

Institutional Review Board Statement: Not applicable.

Informed Consent Statement: Not applicable.

Data Availability Statement: The data presented in this study are available on request from the corresponding author.

Acknowledgments: The authors thank Iris Weber and Maximilian P. Reitze, both from the Institute for Planetology, Westphalian Wilhelms-University, Muenster, Germany, for FTIR spectroscopy.

Conflicts of Interest: The authors declare no conflict of interest.

References

1. Chybowski, E.A.; Glickman, G.N.; Patel, Y.; Fleury, A.; Solomon, E.; He, J. Clinical Outcome of Non-Surgical Root Canal Treatment Using a Single-Cone Technique with Endosequence Bioceramic Sealer: A Retrospective Analysis. *J. Endod.* **2018**, *44*, 941–945. [CrossRef] [PubMed]
2. Zavattini, A.; Knight, A.; Foschi, F.; Mannocci, F. Outcome of Root Canal Treatments Using a New Calcium Silicate Root Canal Sealer: A Non-Randomized Clinical Trial. *J. Clin. Med.* **2020**, *9*, 782. [CrossRef] [PubMed]
3. Camilleri, J.; Atmeh, A.; Li, X.; Meschi, N. Present Status and Future Directions: Hydraulic Materials for Endodontic Use. *Int. Endod. J.* **2022**, *55* (Suppl. S3), 710–777. [CrossRef] [PubMed]
4. Urban, K.; Neuhaus, J.; Donnermeyer, D.; Schäfer, E.; Dammaschke, T. Solubility and PH Value of 3 Different Root Canal Sealers: A Long-Term Investigation. *J. Endod.* **2018**, *44*, 1736–1740. [CrossRef]
5. Donnermeyer, D.; Bürklein, S.; Dammaschke, T.; Schäfer, E. Endodontic Sealers Based on Calcium Silicates: A Systematic Review. *Odontology* **2019**, *107*, 421–436. [CrossRef]
6. Aminoshariae, A.; Primus, C.; Kulild, J.C. Tricalcium Silicate Cement Sealers: Do the Potential Benefits of Bioactivity Justify the Drawbacks? *J. Am. Dent. Assoc.* **2022**, *153*, 750–760. [CrossRef]
7. Donnermeyer, D.; Urban, K.; Bürklein, S.; Schäfer, E. Physico-Chemical Investigation of Endodontic Sealers Exposed to Simulated Intracanal Heat Application: Epoxy Resins and Zinc Oxide–Eugenols. *Int. Endod. J.* **2020**, *53*, 690–697. [CrossRef]
8. *ISO 6876:2012; Dentistry—Root Canal Sealing Materials.* International Organization for Standardization: Geneva, Switzerland, 2012.
9. Donnermeyer, D.; Ibing, M.; Bürklein, S.; Weber, I.; Reitze, M.P.; Schäfer, E. Physico-Chemical Investigation of Endodontic Sealers Exposed to Simulated Intracanal Heat Application: Hydraulic Calcium Silicate-Based Sealers. *Materials* **2021**, *14*, 728. [CrossRef]
10. Donnermeyer, D.; Schäfer, E.; Bürklein, S. Real-Time Intracanal Temperature Measurement During Different Obturation Techniques. *J. Endod.* **2018**, *44*, 1832–1836. [CrossRef]
11. Chen, B.; Haapasalo, M.; Mobuchon, C.; Li, X.; Ma, J.; Shen, Y. Cytotoxicity and the Effect of Temperature on Physical Properties and Chemical Composition of a New Calcium Silicate-Based Root Canal Sealer. *J. Endod.* **2020**, *46*, 531–538. [CrossRef]
12. Ylmén, R.; Jäglid, U.; Steenari, B.M.; Panas, I. Early Hydration and Setting of Portland Cement Monitored by IR, SEM and Vicat Techniques. *Cem. Concr. Res.* **2009**, *39*, 433–439. [CrossRef]
13. Asensio, R.C.; San Andrés Moya, M.; De La Roja, J.M.; Gómez, M. Analytical Characterization of Polymers Used in Conservation and Restoration by ATR-FTIR Spectroscopy. *Anal. Bioanal. Chem.* **2009**, *395*, 2081–2096. [CrossRef]
14. Fawcett, W.R.; Liu, G.; Kessler, T.E. Solvent-Induced Frequency Shifts in the Infrared Spectrum of Acetonitrile in Organic Solvents. *J. Phys. Chem.* **1993**, *97*, 9293–9298. [CrossRef]
15. Donnermeyer, D.; Göbel, L.; Bürklein, S.; Dammaschke, T.; Schäfer, E. Duration of Immersion and Type of Immersion Solution Distort the Outcome of Push-Out Bond Strength Testing Protocols. *Materials* **2019**, *12*, 2860. [CrossRef]
16. Benezra, M.K.; Wismayer, P.S.; Camilleri, J. Influence of Environment on Testing of Hydraulic Sealers. *Sci. Rep.* **2017**, *7*, 17927. [CrossRef]
17. Chedella, S.C.V.; Berzins, D.W. A Differential Scanning Calorimetry Study of the Setting Reaction of MTA. *Int. Endod. J.* **2010**, *43*, 509–518. [CrossRef]
18. Silva, E.J.N.L.; Cardoso, M.L.; Rodrigues, J.P.; De-Deus, G.; Fidalgo, T.K.d.S. Solubility of Bioceramic- and Epoxy Resin-Based Root Canal Sealers: A Systematic Review and Meta-Analysis. *Aust. Endod. J.* **2021**, *47*, 690–702. [CrossRef]
19. Zhou, H.M.; Shen, Y.; Zheng, W.; Li, L.; Zheng, Y.F.; Haapasalo, M. Physical Properties of 5 Root Canal Sealers. *J. Endod.* **2013**, *39*, 1281–1286. [CrossRef]
20. Ersahan, S.; Aydin, C. Solubility and Apical Sealing Characteristics of a New Calcium Silicate-Based Root Canal Sealer in Comparison to Calcium Hydroxide-, Methacrylate Resin- and Epoxy Resin-Based Sealers. *Acta Odontol. Scand.* **2013**, *71*, 857–862. [CrossRef]
21. Borges, R.P.; Sousa-Neto, M.D.; Versiani, M.A.; Rached-Júnior, F.A.; De-Deus, G.; Miranda, C.E.S.; Pécora, J.D. Changes in the Surface of Four Calcium Silicate-Containing Endodontic Materials and an Epoxy Resin-Based Sealer after a Solubility Test. *Int. Endod. J.* **2012**, *45*, 419–428. [CrossRef]
22. Mohammadi, Z.; Dummer, P.M.H. Properties and Applications of Calcium Hydroxide in Endodontics and Dental Traumatology. *Int. Endod. J.* **2011**, *44*, 697–730. [CrossRef]
23. Sanz, J.L.; López-García, S.; Rodríguez-Lozano, F.J.; Melo, M.; Lozano, A.; Llena, C.; Forner, L. Cytocompatibility and Bioactive Potential of AH Plus Bioceramic Sealer: An in Vitro Study. *Int. Endod. J.* **2022**, *55*, 1066–1080. [CrossRef]
24. Aksel, H.; Makowka, S.; Bosaid, F.; Guardian, M.G.; Sarkar, D.; Azim, A.A. Effect of Heat Application on the Physical Properties and Chemical Structure of Calcium Silicate-Based Sealers. *Clin. Oral Investig.* **2021**, *25*, 2717–2725. [CrossRef]
25. Hadis, M.; Camilleri, J. Characterization of Heat Resistant Hydraulic Sealer for Warm Vertical Obturation. *Dent. Mater.* **2020**, *36*, 1183–1189. [CrossRef]

Article

Therapeutic Manuka Honey as an Adjunct to Non-Surgical Periodontal Therapy: A 12-Month Follow-Up, Split-Mouth Pilot Study

David Opšivač [1], Larisa Musić [2], Ana Badovinac [2], Anđelina Šekelja [3] and Darko Božić [2,4,*]

1. School of Medicine, University of Pula, Zagrebačka 30, 52100 Pula, Croatia
2. Department of Periodontology, School of Dental Medicine, University of Zagreb, Gunduliceva 5, 10000 Zagreb, Croatia
3. Health Center Zagreb, Runjaninova 4, 10000 Zagreb, Croatia
4. University Dental Clinic, University Hospital Centre Zagreb, Kišpatićeva 12, 10000 Zagreb, Croatia
* Correspondence: bozic@sfzg.hr; Tel.: +385-14802202

Abstract: Periodontitis is recognized as one of the most common diseases worldwide. Non-surgical periodontal treatment (NSPT) is the initial approach in periodontal treatment. Recently, interest has shifted to various adjunctive treatments to which the bacteria cannot develop resistance, including Manuka honey. This study was designed as a split-mouth clinical trial and included 15 participants with stage III periodontitis. The participants were subjected to non-surgical full-mouth therapy, followed by applying Manuka honey to two quadrants. The benefit of adjunctive use of Manuka honey was assessed at the recall appointment after 3, 6, and 12 months, when periodontal probing depth (PPD), split-mouth plaque score (FMPS), split-mouth bleeding score (FMBS), and clinical attachment level (CAL) were reassessed. Statistically significant differences between NSPT + Manuka and NSPT alone were found in PPD improvement for all follow-up time points and CAL improvement after 3 and 6 months. These statistically significant improvements due to the adjunctive use of Manuka amounted to (mm): 0.21, 0.30, and 0.19 for delta CAL and 0.18, 0.28, and 0.16 for delta PPD values measured after 3, 6, and 12 months, respectively. No significant improvements in FMPS and FMBS were observed. This pilot study demonstrated the promising potential of Manuka honey for use as an adjunct therapy to nonsurgical treatment.

Keywords: periodontitis; manuka honey; nonsurgical periodontal therapy

1. Introduction

Periodontitis is a chronic inflammatory disease affecting the teeth's supporting apparatus. Bacterial biofilm and the associated periodontal pathogenic bacteria, mainly Gram-negative anaerobes, are the main etiological factor of the disease [1].

The main goal of periodontal treatment is to reduce the number of periodontal pathogens and arrest the inflammatory process. The contemporary gold treatment standard is non-surgical periodontal therapy (NSPT), which involves scaling and root planning using manual and machine-driven (sonic or ultrasonic) instruments [2]. The literature suggests that this therapy is highly effective in eliminating the infection. The latest systematic review article by Suvan et al. on subgingival instrumentation for periodontitis treatment estimates a weighted range of pocket depth reduction of 1.0–1.7 mm and a ratio of pocket closure of 57–74% after 3/4 and 6/8 months, respectively, that was achieved through non-surgical periodontal treatment only [3]. Although NSPT can effectively reduce the number of periodontal pathogens, microbial recolonization commonly occurs, and residual pockets are expected to remain after NSPT [2].

Citation: Opšivač, D.; Musić, L.; Badovinac, A.; Šekelja, A.; Božić, D. Therapeutic Manuka Honey as an Adjunct to Non-Surgical Periodontal Therapy: A 12-Month Follow-Up, Split-Mouth Pilot Study. *Materials* **2023**, *16*, 1248. https://doi.org/10.3390/ma16031248

Academic Editor: Lia Rimondini

Received: 10 December 2022
Revised: 15 January 2023
Accepted: 28 January 2023
Published: 1 February 2023

Copyright: © 2023 by the authors. Licensee MDPI, Basel, Switzerland. This article is an open access article distributed under the terms and conditions of the Creative Commons Attribution (CC BY) license (https://creativecommons.org/licenses/by/4.0/).

Various systemically administered and locally delivered adjuncts to NSPT have been suggested, including systemic and local antibiotics, antiseptics, probiotics, lasers, and photodynamic treatment. However, the latest guidelines on the treatment of periodontitis stage I–III do not support the use of adjuncts. The exception in terms of open recommendations is given for locally administered sustained-release chlorhexidine and antibiotics and the use of systemic antibiotics in specific patient groups [2].

The fact that bacteria are becoming increasingly resistant to antibiotics and antiseptics has shifted the interest of medicine to alternative treatment methods against which bacterial resistance cannot be developed. This approach includes using honey, which is increasingly used in medicine. Since the 1990s, when the first studies appeared on the therapeutic effects of honey, particular interest has been focused on its antibacterial properties against infections and antibiotic-resistant bacteria. This effect is consequential mainly of the high sugar concentration of honey, its low pH value, and the formation of hydrogen peroxide that occurs in the enzymatic breakdown of glucose by the glucose oxidase enzyme [4]. Contemporary research on the effects of honey focuses predominantly on one specific honey type, leading to the medicinal use of Manuka honey due to its antibacterial properties [5]. This is an endemic type of honey produced by bees in Australia and New Zealand from the flowers of the plant *Leptospermum scoparium* [6].

The concentration of hydrogen peroxide in Manuka honey is lower than in other types of honey [7]. The specific antibacterial activity in Manuka honey is based on methylglyoxal (MGO), a compound proven to be a very efficient bactericide, virucide, and fungicide. Furthermore, Manuka honey is highly effective against antibiotic-resistant bacteria [8]. The antibacterial potency of Manuka honey was found to be related to its Non-Peroxide Activity (NPA), trademarked as Unique Manuka Factor (UMF) rating, a classification system which reflects the equivalent concentration of phenol (%, w/v) required to produce the same antibacterial activity as honey, and it is correlated with the methylglyoxal and total phenols content [9]. In addition to its antimicrobial properties, published literature suggests that MGO also has immunomodulatory effects which may positively impact wound healing and tissue regeneration [10,11].

Therapeutic Manuka honey has not yet been investigated as a possible adjunct to NSPT. Therefore, this pilot study aims to evaluate the effects of a product containing Manuka honey on periodontal parameters when applied to periodontal pockets after nonsurgical periodontal treatment in patients with stage 3 periodontitis.

2. Materials and Methods

2.1. Experimental Design

This study was designed as a single-center prospective pilot trial with a 12-month follow-up. A split-mouth study model was used. Two quadrants were randomly assigned to the test treatment of NSPT + product containing Manuka honey or NSPT-only.

This pilot study was approved by the Ethics Committee of the School of Dental Medicine, University of Zagreb, Croatia; approval No. 05-PA-30-IX-9/2019. All parts of the study were conducted in full accordance with the World Medical Association Declaration of Helsinki on ethical principles for medical research involving human subjects.

2.2. Population Screening and Inclusion

Patients who sought or were referred for periodontal therapy at the Clinical Department of Periodontology, University Hospital Zagreb, between September 2019 and March 2021, were screened for possible inclusion in the study. The inclusion criteria were: (1) systemically healthy patients of both genders, between the age of 18 and 70; (2) non-smokers; (3) presence of at least 20 teeth; and (4) untreated generalized advanced chronic periodontitis according to the 1999 Classification 1999 [12], i.e., generalized stage III periodontitis according to the 2007 Classification [13]. Exclusion criteria were: (1) pregnant and nursing women; (2) antibiotics prescribed for dental or non-dental diseases six months before the start of the research; (3) systemic diseases or the use of drugs known to affect periodontal tissues; and (4) acute oral or periodontal inflammation or infection (pericoronitis, necrotizing periodontal diseases, etc.). Following inclusion, a periodontal examination was performed by one calibrated periodontist (D.B.). Assessments were done at six sites using a UNC-15 periodontal probe (HuFriedy, Chicago, IL, USA). The following parameters were measured and recorded: probing pocket depth (PPD), recession of the gingival margin (REC), clinical attachment loss (CAL; calculated as the sum of PPD and REC), split-mouth bleeding score (SMBS; calculated as the percentage of positive bleeding sites on probing and expressed for NSPT + Manuka and NSPT-only quadrants, respectively) and split-mouth plaque score (SMPS; calculated as the percentage of sites with present plaque and expressed for NSPT + Manuka and NSPT-only quadrants, respectively) [14,15]. Third molars, if present, were excluded from data analysis.

All participants have given written informed consent to study participation.

2.3. Periodontal Treatment

Nonsurgical treatment was performed by standardized protocol by a single operator (D.O.). All patients received identical oral hygiene instructions, presuming the use of appropriately sized interdental brushes and manual toothbrushes with regular fluoride-containing toothpaste. The use of mouthwashes of any formulation was not allowed during the study period. Mechanical subgingival instrumentation was performed using an ultrasonic instrument (Piezon, E.M.S. Electro Medical Systems S.A., Nyon, Switzerland) and curettes (BioGent, Hu-Friedy, Chicago, IL, USA). Local anesthesia (Ubistesin 40 mg/mL + 0.005 mg/L, 3M Deutschland GmbH, Seefeld, Germany) was provided to all participants. Instrumentation was performed according to the individual situation and without any time limitation. All treatments were concluded within the timeframe of 24 h.

The adjunctive treatment used in this study was a novel commercial product (Pocket Protect, CleverCool B.V., Lijnden, The Netherlands) containing therapeutic Manuka honey and hydrogen peroxide. The two substances mix within a double-barrel syringe before deposition within the pocket.

The product was administered in all pockets (depths $\geq$ 4 mm) in the two active quadrants as per the instructions of the manufacturer. Once the inserted syringe reached the bottom of the periodontal pocket, the product was extruded until the excess was observed in the sulcus. Subjects were not allowed to consume drinks or food for at least 30 min following the procedure.

All subjects were required to report possible adverse effects.

The patients were scheduled for recall visits after three, six, and 12 months when PPD and CAL were re-evaluated. In the first three months only were the patients scheduled for supportive treatment at one-month intervals. The supportive treatment consisted of OH re-instructions, if deemed necessary, and supragingival scaling and polishing. The collected data were pseudo-anonymized immediately after collection. Only the clinician performing the treatment had the access to the patient's identifying information.

2.4. Randomization and Blinding

Randomization of patients was done using a computerized random number generator. Each quadrant was allocated to receive one of the two treatments (Manuka + NSPT or NSPT-only), with the allocation ratio forced to 1:1. The concealment was achieved using sealed and numbered envelopes. A researcher not involved in the operative phases of the study performed the random allocation sequence and intervention assignment.

Blinding was not possible during the experimental period (operator, subjects) due to the specific design of the product-delivery syringe and product's taste.

The examiner was unaware of the treatment allocation at any point during the ongoing study period. Blinding was also done for statistical analysis.

2.5. Statistical Analysis

As the assumption of normality of distribution was verified by inspecting normal Q-Q plots, the comparisons between the NSPT + Manuka and the NSPT-only quadrants were performed using a two-tailed t-test for independent samples with the assumption of homoskedasticity. PPD and CAL values were compared between the NSPT + Manuka and the NSPT-only quadrants at each time point (baseline, three months, six months, and 12 months). The changes in the parameters PPD and CAL (denoted as delta PPD and delta CAL, respectively) were calculated for each time point by subtracting the baseline values from the values measured after three months, six months, and 12 months. The obtained delta values were statistically compared between the NSPT + Manuka and the NSPT-only quadrants using a two-tailed t-test for independent samples. BoP index and plaque index were represented as percentages of sites that were positive for bleeding or the presence of plaque, respectively. These percentages of positive sites were compared between the NSPT + Manuka and the NSPT-only quadrants using the chi-square test.

The statistical analysis was performed using SPSS (version 25; IBM, Armonk, NY, USA) at a level of significance of 0.05.

3. Results

A total of 86 patients were screened for inclusion and 15 participants (eight males and seven female) were recruited for this pilot study. However, three of them were excluded from the study due to not showing up at the follow-up appointment at 3 months (two participants) and 6 months (one participant). Hence, a total of 12 participants (five males and seven female) were included. The mean age was 43.1 years (range 31–49), with a mean number of teeth at 26, and 1331 sites with increased PPD ($\geq$4 mm); 4–6 mm at 905 sites and >6 mm at 426 sites. All patients were non-smokers.

Table 1 shows the PPD and CAL values, and the corresponding delta values representing differences from baseline measured after 3, 6, and 12 months. There were no statistically significant differences in baseline values for PPD and CAL between the sites treated with NSPT + Manuka and the sites treated with NSPT-only. In addition, no significant differences were observed for PPD and CAL in the comparisons with the sites treated with NSPT + Manuka and the sites treated with NSPT-only at each of the follow-up time periods (3, 6, and 12 months). However, statistically significant differences between NSPT + Manuka and NSPT-only were identified in delta PPD values for all follow-up time points, as well as for delta CAL values for the time points of 3 and 6 months. The delta CAL values calculated for 12 months could be considered marginally significant at the selected level of significance of 0.05. These statistically significant further improvements due to the adjunctive use of Manuka amounted to (mm): 0.21, 0.30, and 0.19 for delta CAL and 0.18, 0.28, and 0.16 for delta PPD values measured after 3, 6, and 12 months, compared to NSPT-only.

Table 2 shows FMPS and FMBS values measured at baseline, 3, 6, and 12 months. At baseline, the quadrants that received NSPT-only had significantly greater values of plaque and bleeding. No significant differences between the quadrant groups were identified, with the exception of lower bleeding scores at 6 months in the NSPT + Manuka quadrants.

One patient reported generalized dentine hypersensitivity that spontaneously decreased until the 1-month check-up. No other adverse effects were reported.

Table 1. Measured values of periodontal pocket depth, clinical attachment level, and changes of these parameters from the baseline values. All data are represented as mean values with 95% confidence interval limits in parentheses.

Variable	Time Point	NSPT + MANUKA	NSPT-Only	p-Values
PPD (mm)	baseline	4.27 (4.12, 4.43)	4.07 (3.93, 4.21)	0.060
	3 months	2.81 (2.72, 2.91)	2.82 (2.73, 2.92)	0.846
	6 months	2.57 (2.48, 2.65)	2.66 (2.57, 2.74)	0.140
	12 months	2.53 (2.45, 2.61)	2.52 (2.44, 2.60)	0.825
ΔPPD (mm)	3 months	−1.46 (−1.56, −1.36)	−1.25 (−1.34, −1.15)	**0.002**
	6 months	−1.71 (−1.81, −1.60)	−1.41 (−1.51, −1.31)	**<0.001**
	12 months	−1.74 (−1.85, −1.63)	−1.55 (−1.66, −1.44)	**0.016**
CAL (mm)	baseline	4.30 (4.14, 4.45)	4.11 (3.97, 4.25)	0.080
	3 months	2.91 (2.81, 3.02)	2.91 (2.81, 3.01)	0.952
	6 months	2.69 (2.60, 2.79)	2.78 (2.68, 2.88)	0.201
	12 months	2.66 (2.57, 2.75)	2.63 (2.54, 2.72)	0.609
ΔCAL (mm)	3 months	−1.38 (−1.49, −1.28)	−1.20 (−1.30, −1.10)	**0.012**
	6 months	−1.61 (−1.72, −1.49)	−1.33 (−1.43, −1.22)	**<0.001**
	12 months	−1.64 (−1.72, −1.52)	−1.48 (−1.59, −1.37)	**0.052**

PPD—periodontal pocket depth; CAL—clinical attachment level; Δ—delta.

Table 2. Measured values of split-mouth bleeding score and split-mouth plaque score, represented as percentages of positive sites in respective quadrants (%).

Variable	Time Point	NSPT + MANUKA	NSPT-Only	p-Values
SMBS (%)	baseline	82.1	86.5	**0.007**
	3 months	46.7	43.7	0.178
	6 months	44.8	50.8	**0.007**
	12 months	54.6	55.8	0.575
SMPS (%)	baseline	92.7	95.4	0.009
	3 months	30.9	31.2	**0.990**
	6 months	46.7	47.4	0.775
	12 months	51.3	52.2	0.678

SMBS—split-mouth bleeding score (number of positive sites/all measuring sites × 100 in Manuka or NSPT-only quadrants); SMPS—split-mouth plaque score (number of positive sites/all measuring sites × 100 in Manuka or NSPT-only quadrants); %—percentage.

4. Discussion

This pilot study investigated the impact of Manuka honey as an adjunct to NSPT. It showed statistically significant improvements in terms of PPD reduction and CAL gain after 3, 6, and 12 months of follow-up, compared to the outcomes of the NSPT-only. Improvements in bleeding and plaque scores were also observed in both quadrant groups.

The present study was motivated by numerous literature reports on the beneficial effects of Manuka honey when used for various medical purposes. Accelerated wound healing was reported after the topical use of Manuka honey for the treatment of ulcers, bed sores, and other skin infections [4]. In addition, Manuka honey was demonstrated to promote healing in infected wounds that do not respond to conventional therapy, i.e., antibiotics and antiseptics, including wounds infected with methicillin-resistant *S. aureus* [16]. Manuka honey may promote the repair of the damaged intestinal mucosa, stimulate the growth of new tissues, and work as an anti-inflammatory agent [16]. Clinical observations have reported reduced symptoms of inflammation when Manuka honey is applied to wounds [17]. The removal of exudate in wounds dressed with honey was helpful for managing inflamed wounds [4]. The aforementioned effects encouraged our investigation of the possible effects of Manuka honey on periodontitis treatment.

Although the effects of Manuka honey on oral bacteria have not been investigated in vivo, there are several in vitro studies that convincingly indicated its antibacterial activity. The study by Safii et al. was based on the minimum amount of honey required to kill bacteria or inhibit their growth on blood agar, as evaluated by the minimum bactericidal concentration or minimum inhibitory concentration (MBC/MIC), and showed a high antibacterial potential of Manuka honey, especially against Gram-negative anaerobic bacteria [18]. Similar results were published in the in vitro work of Schmidlin et al. on the antimicrobial effect of Manuka honey compared to other types of honey against three common pathogens present in the oral cavity (*S. mutans*, *P. gingivalis*, and *A. actinomycetemcommitans*). Antibacterial activity was analyzed on blood agar, and Manuka honey demonstrated a stronger antibacterial effect compared to other types of honey due to its specific non-peroxidase activity mediated by methylglyoxal, to which *P. gingivalis* was found to be especially sensitive [17].

The research by Badet and Quero analyzed in vitro adhesion of oral cavity bacteria exposed to different concentrations of Manuka honey on glass and hydroxyapatite surfaces immersed in saliva. The results showed that Manuka honey with higher methylglyoxal concentrations inhibits the adhesion of *S. mutans* [19]. Sigrun et al. compared the effectiveness of Manuka honey and regular honey on *P. gingivalis*, which is one of the most important periodontal pathogenic bacteria. It was shown that the planktonic form is extremely sensitive to exposure to Manuka honey, which additionally substantiates its antibacterial properties [20].

The only in vivo study on the use of Manuka honey in the oral cavity is reported in the article by Al-Khanati et al. who investigated the analgesic effect of Manuka honey applied to the post-extraction alveolus following the extraction of impacted third molars. Their randomized split-mouth study showed a statistically significant reduction in postoperative pain on the side where Manuka honey was applied [21]. This finding suggests favorable analgesic properties and potential anti-inflammatory activity of Manuka honey.

A recent systematic review and meta-analysis analyzed the adjunctive effects of locally delivered antimicrobials in the nonsurgical treatment of periodontitis. The meta-analysis for studies of 6–9-month follow-ups showed statistically significant benefits in terms of PPD reduction (WMD = 0.365) and CAL gain (WMD = 0.263). For long-term studies of 12 to 60 months, significant differences were only observed for PPD (WMD = 0.190). The authors highlighted that the heterogeneity was significant due to a great number of different active agents and a difference in study design. The products with the largest observed benefits were those containing doxycycline or tetracycline [22]. The improvements in the treatment outcome following the adjunctive use of Manuka honey in the present study can be compared to the systematic review's results in terms of PPD reduction, as the mean difference at 6 and 12 months were 0.30 and 0.19, respectively.

Although the differences between the two quadrant groups in our study seem small and that the baseline values might also seem low, it is interesting to note that in a recent multicenter clinical trial investigating the flapless application of enamel matrix derivative in non-surgical therapy when the authors sub-analyzed CAL changes at 12 months including all pockets, and not only the deep ones, the difference between the control and test group was only 0.1 mm in favor of the enamel matrix derivative, with overall CAL changes of 0.8 mm and 0.9 mm, respectively. However, when they analyzed only pockets of 5–8 mm, CAL changes were significantly higher, 2.1 mm for the control group and 2.2 mm for the test group [23]. If we compare the results from the above-mentioned study on all sites to ours, we can see that our CAL changes in the test quadrants were 1.64 mm and 1.48 mm in the control quadrants which is higher than in the Schallhorn study.

Another observation in our study was the continuous improvements in both PPD reductions and CAL values at 3, 6, and 12 months. However, statistical significance between the two groups was only found at 6 months in favor of the test quadrants for both variables. At 3 months, PPD changes were −1.46 mm for the test quadrants and −1.25 mm coming very close to statistical significance (p = 0.002) with a further improvement to −1.74 mm in test quadrants and −1.55 mm in the control quadrants at 12 months with p values still not significant. This difference in the lack of significance at 3 and 12 months compared to 6 months, although the actual numbers in differences are 0.21 mm, 0.30 mm, and 0.19 mm, could be due to the pressure during probing which cannot be entirely controlled although clinical measurements were done by an experienced and calibrated periodontist.

Both quadrant groups presented with an initial reduction in plaque and bleeding scores following treatment. After 3 months, at re-evaluation, the reduction of bleeding was 35.4% in the Manuka + NSPT and 42.8% in NSPT-only quadrants. These values are lower than previously reported by the systematic review of Suvan et al., of a weighted mean reduction of 56.7% at 3/4 months after non-surgical treatment [3]. At 6 months, while the Manuka + NSPT quadrants presented a further, albeit minimal, reduction of 1.9% in bleeding scores, the NSPT-only quadrants presented a rise in bleeding scores of 7.1%. From 6 to 12 months, quadrants treated with both treatment modalities presented with an increase in bleeding scores. The observed reduction of plaque (SMPS) at 3 months was 61.8% in the Manuka + NSPT quadrants and 64.2% in the NSPT-only quadrants, which is as expected following nonsurgical treatment [24]. At 6 and 12 months, quadrants treated with both modalities expressed an increase in plaque score. The increase in both parameters at later follow-ups could be explained by the omission of intensive supportive treatment at 1-month intervals and lack of patients' motivation. It needs to be highlighted that the data on the effects of these two treatment modalities and the direct comparison between the two treatments on plaque and bleeding have to be interpreted with caution, as there was a statistically significant difference in these parameters between the Manuka + NSPT and NSPT-only quadrants at baseline. As per the effects of Manuka honey on plaque and bleeding, a pilot study by Molan et al. showed that Manuka honey has a positive effect on reducing the amount of dental plaque and the incidence of gingivitis compared to the control group that did not use Manuka honey [7].

The commercial product used in this research combines Manuka honey and hydrogen peroxide. Thus, the effect of the adjunctive cannot be attributed solely to Manuka. The extent of these substances' sole or combined impact on clinical outcomes would need to be evaluated in an appropriately designed RCT. Schmidlin et al., however, reported a greater antibacterial effect of Manuka above the NPA value of 15 compared to "regular" honey, whose activity relies on peroxide-based antimicrobial factors [17]. Hydrogen peroxide as an adjunct to non-surgical treatment for pocket irrigation failed to show further improvements in clinical outcomes when compared to subgingival debridement alone [24]. Research on the use of 1.7% hydrogen peroxide in custom trays and H_2O_2 photolysis (irradiation of 3% H_2O_2 with 405 nm light or diode laser) showed promising results in terms of significantly greater PPD reduction in test groups compared to controls [25–28].

Adding hydrogen peroxide in the commercial product definitely facilitates clinical handling as it decreases the honey's viscosity and the application inside the periodontal pockets.

The main limitation of this pilot study is the small sample size (n = 12). However, despite the small sample size, the included subjects represent a specific population of patients with severe inflammation and advanced forms of periodontitis, for which it was possible to identify a small but statistically significant improvement in the reduction of PPD at sites treated with Manuka honey compared to the control sites. An analogous improvement was identified for CAL in the sites treated with Manuka honey.

Furthermore, microbiological analysis was not conducted in this research. So far, antibacterial efficacy of Manuka on periodontal pathogens was shown only in in vitro studies [17,18]. Thus, further data is needed from studies on humans. Studies with mid-

and long-term follow-ups may show possible prolonged effects of this adjunctive on the subgingival microbial composition compared to subgingival debridement alone.

The statistical analysis was performed at the level of individual periodontal sites, meaning that the statistical unit was represented by the site instead of the patient. When individual sites are considered as statistical units, the actual sample size becomes 2025 (12 subjects × 28.125 teeth per subject on average × 6 sites per tooth = 2025 sites in total). Calculating the "delta" values for PPD and CAL and using these values instead of the original pooled PPD and CAL values measured at each individual time point provided higher statistical power for identifying significant differences in the comparisons of the NSPT + Manuka and the NSPT-only quadrants. By calculating the differences in PPD and CAL at the level of the individual site, the heterogeneity among sites becomes less influential, as the "net" effect is isolated for each site and used for the statistical analysis rather than pooling the original PPD and CAL values measured at individual time points. This reasoning explains why the significant effects of Manuka honey were identified only for delta PPD and delta CAL variables and not for the original PPD and CAL variables. Furthermore, the authors are aware of the possible disadvantages of the split-mouth design. As both sides received some modality of treatment and the product itself comes in a highly viscous form and is delivered locally, the possibility of the carry-across effect (i.e., bacterial contamination from an untreated sides/possible effect of the Manuka honey on the NSPT-only quadrants) was minimized [29,30]; however, we do recognize a risk of it happening. A relatively high number of teeth (20 teeth) set as an inclusion criterion also positively influenced the similarity between randomization units.

5. Conclusions

This pilot study indicated a promising potential of Manuka honey as an adjunct in NSPT. Despite the improvement in outcomes appearing modest in absolute terms, it was statistically significant for all follow-up time points, indicating the potential use of Manuka honey as a simple and affordable adjunct to non-surgical periodontal therapy. Furthermore, the product is considered a safe adjunct and no adverse events related to its use were reported during the study period. Encouraging results from this pilot study led to a further randomized clinical study on a larger sample that is currently being performed by our group.

Author Contributions: D.B.—study conception, experimental design, funding acquisition, performed experiments, formal analysis and data interpretation, statistical analysis, visualization, and wrote manuscript; D.O., L.M., A.B., and A.Š.—experimental design, performed experiments, data acquisition and interpretation, and revised and approved final manuscript; D.B.—study conception, experimental design, supervision, data interpretation, project administration, resources, and revised and approved final manuscript. D.B.—experimental design, data acquisition and interpretation, supervision, resources, and revised and approved final manuscript. D.O. and L.M.—study conception, experimental design, supervision, data interpretation, project administration, resources, and revised and approved final manuscript. D.B.—study conception, supervision, resources, data interpretation, project administration, and revised and approved final manuscript. All authors have read and agreed to the published version of the manuscript.

Funding: This research received no external funding.

Institutional Review Board Statement: The study was conducted in accordance with the Declaration of Helsinki, and approved by the Ethics Committee of the School of Dental Medicine, University of Zagreb, Croatia; approval No. 05-PA-30-IX-9/2019.

Informed Consent Statement: Informed consent was obtained from all subjects involved in the study.

Data Availability Statement: Not applicable.

Conflicts of Interest: The authors declare no conflict of interest.

References

1. Chapple, I.L.C.; Mealey, B.L.; Van Dyke, T.E.; Bartold, P.M.; Dommisch, H.; Eickholz, P.; Geisinger, M.L.; Genco, R.J.; Glogauer, M.; Goldstein, M.; et al. Periodontal Health and Gingival Diseases and Conditions on an Intact and a Reduced Periodontium: Consensus Report of Workgroup 1 of the 2017 World Workshop on the Classification of Periodontal and Peri-Implant Diseases and Conditions. *J. Periodontol.* **2018**, *89*, S74–S84. [CrossRef]
2. Sanz, M.; Herrera, D.; Kebschull, M.; Chapple, I.; Jepsen, S.; Berglundh, T.; Sculean, A.; Tonetti, M.S.; EFP Workshop Participants and Methodological Consultants. Treatment of Stage I–III Periodontitis—The EFP S3 Level Clinical Practice Guideline. *J. Clin. Periodontol.* **2020**, *47*, 4–60. [CrossRef] [PubMed]
3. Suvan, J.; Leira, Y.; Moreno Sancho, F.M.; Graziani, F.; Derks, J.; Tomasi, C. Subgingival Instrumentation for Treatment of Periodontitis. A Systematic Review. *J. Clin. Periodontol.* **2020**, *47*, 155–175. [CrossRef] [PubMed]
4. Carter, D.A.; Blair, S.E.; Cokcetin, N.N.; Bouzo, D.; Brooks, P.; Schothauer, R.; Harry, E.J. Therapeutic Manuka Honey: No Longer So Alternative. *Front. Microbiol.* **2016**, *7*, 569. [CrossRef] [PubMed]
5. Maddocks, S.E.; Jenkins, R.E. Honey: A Sweet Solution to the Growing Problem of Antimicrobial Resistance? *Future Microbiol.* **2013**, *8*, 1419–1429. [CrossRef] [PubMed]
6. Mandal, M.D.; Mandal, S. Honey: Its Medicinal Property and Antibacterial Activity. *Asian Pac. J. Trop. Biomed.* **2011**, *1*, 154–160. [CrossRef] [PubMed]
7. Adams, C.J.; Manley-Harris, M.; Molan, P.C. The Origin of Methylglyoxal in New Zealand Manuka (*Leptospermum Scoparium*) Honey. *Carbohydr. Res.* **2009**, *344*, 1050–1053. [CrossRef]
8. Norton, A.M.; McKenzie, L.N.; Brooks, P.R.; Pappalardo, L.J. Quantitation of Dihydroxyacetone in Australian Leptospermum Nectar via High-Performance Liquid Chromatography. *J. Agric. Food Chem.* **2015**, *63*, 6513–6517. [CrossRef]
9. Alvarez-Suarez, J.; Gasparrini, M.; Forbes-Hernández, T.; Mazzoni, L.; Giampieri, F. The Composition and Biological Activity of Honey: A Focus on Manuka Honey. *Foods* **2014**, *3*, 420–432. [CrossRef]
10. Tashkandi, H. Honey in Wound Healing: An Updated Review. *Open Life Sci.* **2021**, *16*, 1091–1100. [CrossRef]
11. Minden-Birkenmaier, B.; Bowlin, G. Honey-Based Templates in Wound Healing and Tissue Engineering. *Bioengineering* **2018**, *5*, 46. [CrossRef] [PubMed]
12. Armitage, G.C. Development of a Classification System for Periodontal Diseases and Conditions. *Ann. Periodontol.* **1999**, *4*, 1–6. [CrossRef] [PubMed]
13. Tonetti, M.S.; Greenwell, H.; Kornman, K.S. Staging and Grading of Periodontitis: Framework and Proposal of a New Classification and Case Definition. *J. Clin. Periodontol.* **2018**, *45*, S149–S161. [CrossRef] [PubMed]
14. Ainamo, J.; Bay, I. Problems and Proposals for Recording Gingivitis and Plaque. *Int. Dent. J.* **1975**, *25*, 229–235.
15. O'Leary, T.J.; Drake, R.B.; Naylor, J.E. The Plaque Control Record. *J. Periodontol.* **1972**, *43*, 38. [CrossRef] [PubMed]
16. Niaz, K.; Maqbool, F.; Bahadar, H.; Abdollahi, M. Health Benefits of Manuka Honey as an Essential Constituent for Tissue Regeneration. *Curr. Drug Metab.* **2017**, *18*, 881–892. [CrossRef] [PubMed]
17. Schmidlin, P.R.; English, H.; Duncan, W.; Belibasakis, G.N.; Thurnheer, T. Antibacterial Potential of Manuka Honey against Three Oral Bacteria in Vitro. *Swiss Dent. J.* **2014**, *124*, 922–924.
18. Safii, S.H.; Tompkins, G.R.; Duncan, W.J. Periodontal Application of Manuka Honey: Antimicrobial and Demineralising Effects In Vitro. *Int. J. Dent.* **2017**, *2017*, 9874535. [CrossRef]
19. Badet, C.; Quero, F. The in Vitro Effect of Manuka Honeys on Growth and Adherence of Oral Bacteria. *Anaerobe* **2011**, *17*, 19–22. [CrossRef]
20. Eick, S.; Schäfer, G.; Kwieciński, J.; Atrott, J.; Henle, T.; Pfister, W. Hone—A Potential Agent against Porphyromonas Gingivalis: An in Vitro Study. *BMC Oral Health* **2014**, *14*, 24. [CrossRef]
21. Al-Khanati, N.M.; Al-Moudallal, Y. Effect of Intrasocket Application of Manuka Honey on Postsurgical Pain of Impacted Mandibular Third Molars Surgery: Split-Mouth Randomized Controlled Trial. *J. Maxillofac. Oral Surg.* **2019**, *18*, 147–152. [CrossRef]
22. Herrera, D.; Matesanz, P.; Martín, C.; Oud, V.; Feres, M.; Teughels, W. Adjunctive Effect of Locally Delivered Antimicrobials in Periodontitis Therapy: A Systematic Review and Meta-analysis. *J. Clin. Periodontol.* **2020**, *47*, 239–256. [CrossRef] [PubMed]
23. Schallhorn, R.A.; McClain, P.K.; Benhamou, V.; Doobrow, J.H.; Grandin, H.M.; Kasaj, A. Application of Enamel Matrix Derivative in Conjunction with Non-surgical Therapy for Treatment of Moderate to Severe Periodontitis: A 12-month, Randomized Prospective, Multicenter Study. *J. Periodontol.* **2021**, *92*, 619–628. [CrossRef]
24. Wennstrom, J.L.; Tomasi, C.; Bertelle, A.; Dellasega, E. Full-Mouth Ultrasonic Debridement versus Quadrant Scaling and Root Planing as an Initial Approach in the Treatment of Chronic Periodontitis. *J. Clin. Periodontol.* **2005**, *32*, 851–859. [CrossRef] [PubMed]
25. Putt, M.S.; Mallatt, M.E.; Messmann, L.L.; Proskin, H.M. A 6-Month Clinical Investigation of Custom Tray Application of Peroxide Gel with or without Doxycycline as Adjuncts to Scaling and Root Planing for Treatment of Periodontitis. *Am. J. Dent.* **2014**, *27*, 273–284. [PubMed]
26. Putt, M.S.; Proskin, H.M. Custom Tray Application of Peroxide Gel as an Adjunct to Scaling and Root Planing in the Treatment of Periodontitis: Results of a Randomized Controlled Trial after Six Months. *J. Clin. Dent.* **2013**, *24*, 100–107. [PubMed]

27. Kanno, T.; Nakamura, K.; Ishiyama, K.; Yamada, Y.; Shirato, M.; Niwano, Y.; Kayaba, C.; Ikeda, K.; Takagi, A.; Yamaguchi, T.; et al. Adjunctive Antimicrobial Chemotherapy Based on Hydrogen Peroxide Photolysis for Non-Surgical Treatment of Moderate to Severe Periodontitis: A Randomized Controlled Trial. *Sci. Rep.* **2017**, *7*, 12247. [CrossRef] [PubMed]
28. Nammour, S.; El Mobadder, M.; Maalouf, E.; Namour, M.; Namour, A.; Rey, G.; Matamba, P.; Matys, J.; Zeinoun, T.; Grzech-Leśniak, K. Clinical Evaluation of Diode (980 Nm) Laser-Assisted Nonsurgical Periodontal Pocket Therapy: A Randomized Comparative Clinical Trial and Bacteriological Study. *Photobiomodulation Photomed. Laser Surg.* **2021**, *39*, 10–22. [CrossRef]
29. Hujoel, P.P.; DeRouen, T.A. A Survey of Endpoint Characteristics in Periodontal Clinical Trials Published 1988-1992, and Implications for Future Studies. *J. Clin. Periodontol.* **1995**, *22*, 397–407. [CrossRef]
30. Lesaffre, E.; Philstrom, B.; Needleman, I.; Worthington, H. The Design and Analysis of Split-Mouth Studies: What Statisticians and Clinicians Should Know: THE SPLIT-MOUTH DESIGN. *Statist. Med.* **2009**, *28*, 3470–3482. [CrossRef]

Disclaimer/Publisher's Note: The statements, opinions and data contained in all publications are solely those of the individual author(s) and contributor(s) and not of MDPI and/or the editor(s). MDPI and/or the editor(s) disclaim responsibility for any injury to people or property resulting from any ideas, methods, instructions or products referred to in the content.

Article

Effects of Ion-Releasing Materials on Dentine: Analysis of Microhardness, Appearance, and Chemical Composition

Ivan Šalinović [1,*], Falk Schwendicke [2], Haitham Askar [3], Jamila Yassine [4] and Ivana Miletić [1]

[1] Department for Endodontics and Restorative Dentistry, School of Dental Medicine, University of Zagreb, Gundulićeva 5, 10000 Zagreb, Croatia; miletic@sfzg.hr
[2] Department of Oral Diagnostics, Digital Health and Health Services Research, Charité—Universitätsmedizin Berlin, Aßmannshauser Straße 4-6, 14197 Berlin, Germany; falk.schwendicke@charite.de
[3] Department of Operative, Preventive and Pediatric Dentistry, Charité—Universitätsmedizin Berlin, Aßmannshauser Straße 4-6, 14197 Berlin, Germany; haitham.askar@charite.de
[4] Department for Dental Prosthetics, Geriatric Dentistry and Functional Theory, Charité—Universitätsmedizin Berlin, Aßmannshauser Straße 4-6, 14197 Berlin, Germany; jamila.yassine@charite.de
* Correspondence: isalinovic@sfzg.hr; Tel.: +385-14802126

Abstract: The aim of this study was to compare the potential of standard ion-releasing materials to repair demineralized lesions with recently introduced alkasite and glass hybrid materials. Glass ionomer (GC Fuji TRIAGE), two glass hybrids (EQUIA Forte HT, Riva SC), calcium silicate cement (Biodentine) and an alkasite (Cention Forte) were tested. A total of 72 human third molars were used for sample preparation; on the dentine surface, a class-I cavity was prepared, and one half was covered with nail varnish. The teeth were subjected to a demineralization protocol, filled with the examined materials, and cut in half. The evaluation included a dentine microhardness assessment ($n = 10$) and SEM/EDS analysis ($n = 2$). The results were analyzed using SPSS 22.0 statistical software and compared using an analysis of variance and Scheffe post-hoc test. The statistical significance level was set to 0.05. Mean microhardness values (HV0.1) after 14 and 28 days were, respectively: EQUIA Forte HT (26.7 ± 1.45 and 37.74 ± 1.56), Riva Self Cure (19.66 ± 1.02 and 29.58 ± 1.18), Cention Forte (19.01 ± 1.24 and 27.93 ± 1.33), Biodentine (23.35 ± 1.23 and 29.92 ± 1.02), GC Fuji TRIAGE (25.94 ± 1.35 and 33.87 ± 5.57) and control group (15.57 ± 0.68 and 15.64 ± 0.82). The results were significantly different between most groups ($p < 0.001$). SEM/EDS revealed varying patterns, material deposits and distinct elemental variations. To conclude, all materials increased microhardness and affected the dentine surface appearance and chemical composition; EQUIA Forte HT demonstrated the most pronounced effects.

Keywords: dentine; ion-releasing materials; microhardness; glass-ionomer cements; glass hybrids; alkasite

Citation: Šalinović, I.; Schwendicke, F.; Askar, H.; Yassine, J.; Miletić, I. Effects of Ion-Releasing Materials on Dentine: Analysis of Microhardness, Appearance, and Chemical Composition. *Materials* 2023, 16, 7310. https://doi.org/10.3390/ma16237310

Academic Editor: Nikolaos Silikas

Received: 16 October 2023
Revised: 13 November 2023
Accepted: 22 November 2023
Published: 24 November 2023

Copyright: © 2023 by the authors. Licensee MDPI, Basel, Switzerland. This article is an open access article distributed under the terms and conditions of the Creative Commons Attribution (CC BY) license (https://creativecommons.org/licenses/by/4.0/).

1. Introduction

Dentine formation is a complex process. It results in different types of dentine, each with its own characteristics [1]. Unlike enamel, dentine has a higher proportion of organic content, around 20%, most of which is collagen [2]. This complex organic matrix is what makes dentine remineralization a challenging process [3]. Demineralization of dentine, usually associated with the progression of a carious lesion, is initiated by a drop in pH. On the other hand, dentine remineralization requires the harmonious reparation of collagen and inorganic apatite, resulting in the intrafibrillar mineralization of collagen [4]. Furthermore, to restore the mechanical features of dentine, the processes of demineralization and remineralization need to be in synergistic connection to enable a precise mineral precipitation, both within the collagen intrafibrillar and interfibrillar spaces [5,6].

The contemporary approach to caries removal favors the usage of methods and materials that preserve hard dental tissue and promote its reparation, minimizing the risk of unintended pulp chamber exposure [7,8]. Additionally, previous research has shown that tissue preservation significantly improves the longevity of restorations [9]. Therefore, a growing emphasis has been placed on ion-releasing materials that elicit a specific biological response and bond to hard dental tissue, thus leading to tissue replacement that can reduce the susceptibility of tooth minerals to dissolution and/or is capable of restoring its attributes [6]. That group includes many different materials with various mechanisms of action. However, all of them are based on delivering ions into mineral-deprived lesions, aiming to repair them. Standard materials in this category are glass ionomer (GIC) and calcium silicate cements. Initially introduced in the 1970s [10], GICs have since been used in different areas of dental medicine due to their properties, which include biocompatibility, bioactivity, and fluoride release, making them efficient in promoting tissue repair and caries prevention [11]. However, their poor mechanical features and low resistance to wear and erosion have prevented them from becoming a long-term restorative material in permanent dentition [12–15]. To overcome these issues, glass hybrids have been introduced; they are reinforced by adding more reactive, smaller silica particle, and a higher molecular-weight acrylic and acid molecule, all of which increases the matrix cross-linking and improves their mechanical properties [16]. In addition, such materials also come with a resin-based coat, further improving their durability.

Before quick-setting modifications were developed, calcium silicate-based cements have mostly been used in endodontics, such as for the treatment of perforated roots, due to their high biocompatibility and ability to promote the formation of minerals [17]. Nowadays, some tri-calcium silicate-based cements are used as restorative materials and dentine replacements, as they induce mineral precipitation dentine formation [6]. While most of the previously listed materials are self-cured, alkasites, a group of recently introduced materials, are dual-cured. They contain an alkaline filler, thus releasing acid-neutralizing ions, as well as fluorides, hydroxyl ions and calcium in an acidic oral cavity environment [18].

Previous research has shown that the mechanical properties of dentin are best repaired when minerals are incorporated into the collagen fibrils, as the mere deposition of minerals into the demineralized lesion does not guarantee functional remineralization [19,20]. The present study assesses the biomechanical properties of dentine by testing microhardness and using scanning electron microscopy in conjunction with energy-dispersive X-ray spectroscopy (SEM/EDS).

We aimed to compare the potential of standard ion-releasing materials (glass ionomer and calcium silicate cements) to repair demineralized lesions with recently introduced alkasite and glass hybrids.

The null hypotheses were:

1. There is no difference in dentine microhardness values among the tested materials.
2. There are no differences in the mineral composition of the specimens treated with the tested materials.
3. There are no differences in the micro-surface appearance between groups.

2. Materials and Methods

The Ethics Committee of the School of Dental Medicine, University of Zagreb approved the protocol for the current study (05-PA-30-III-12/2021). The research was conducted at the University of Zagreb, School of Dental Medicine, the Ruđer Bošković Institute in Zagreb, the Faculty of Mechanical Engineering and Naval Architecture, University of Zagreb, and the Charite—Universitätsmedizin in Berlin. A total of 72 human third molars that were used in this study were collected at the Department of Oral Surgery, Clinical Hospital Centre Zagreb. Extracted teeth were thoroughly examined to ensure the absence of any carious lesions. Before use, the samples were placed in 0.5% chloramine solution at room temperature for up to three months.

2.1. Sample Preparation

Before the sample preparation, the teeth were meticulously cleaned using brushes, a scaler, and discs. The tested materials are listed in Table 1 (the type of material is provided by the manufacturer).

Table 1. Materials used in the study.

Material	Type	Manufacturer
EQUIA Forte® HT	Glass hybrid	GC Corporation, Tokyo, Japan
Riva Self Cure	Glass hybrid/glass ionomer cement [21]	SDI Limited, Melbourne, VI, Australia
Cention Forte	Alkasite	Ivoclar Vivadent AG, Schaan, Liechenstein
Biodentine™	Tricalcium silicate-based material	SEPTODONT, Saint-Maur-des-fossés Cedex, France
GC Fuji TRIAGE®	Glass ionomer cement	GC Corporation, Tokyo, Japan

A non-bioactive composite material, 3M™ Filtek™ Universal Composite (3M ESPE, St. Paul, MN, USA), was used in the control group. The collected teeth were randomly divided into two groups for each of the five materials tested and a control group, as each of the two tests (after 14 and 28 days) used different specimens. Ten samples were obtained for each group in a single testing period for microhardness testing ($n = 10$). Two additional samples were prepared for SEM/EDS analysis in each group ($n = 2$). The occlusal third of the crown was removed using IsoMet 1000 Precision Cutter (Buehler, Lake Bluff, IL, USA) and IsoMet Diamond Wafering Blade (Buehler, Lake Bluff, IL, USA) at a speed of 200 rounds per minute, exposing the dentine surface. A class-I cavity with the floor ending at mid-coronal dentin (3 mm × 1.5 mm wide, 0.5 mm deep) was prepared in each tooth using a medium-grit (107 m) diamond bur (Komet, Lemgo, Germany) fixed in a water-cooled high-speed turbine. Half of the cavity was covered with acid-resistant nail polish (Markwins Beauty Brands, Inc., Walnut, CA, USA) to enable a direct comparison of surfaces. The samples were then demineralized by individually immersing them in a solution containing 0.0476 mM sodium fluoride (NaF), 2.2 mM calcium chloride dihydrate ($CaCl_2 \cdot 2H_2O$), 2.2 mM potassium dihydrogen phosphate (KH_2PO_4), 50 mM acetic acid (CH_3COOH), and 10 mM potassium hydroxide (KOH) at pH 5.0 (37 °C) for two weeks in an incubator ES 120 (NÜVE, Ankara, Turkey), as suggested by previous studies [22,23]. The cavities were then rinsed and air-dried, after which they were filled with one of the materials studied. All the materials were in encapsulated form and were mixed according to the manufacturer's instructions in the Silver Mix capsule mixer (GC Corporation Tokyo, Japan); those treated with Cention Forte were further cured for 40 s with the light cure unit Woodpecker LED-C (Guilin Tucano Medical Apparatus and Instruments Limited Company, Guilin, China), curing light output: 850 W/cm^2 wavelength: 420–480 nm. The samples were placed in saline (Croatian Institute of Transfusion Medicine, Zagreb, Croatia) mixed with the same amount of oral cavity moisturizer (Certmedica International GmbH, Aschaffenburg, Germany) for 14 and 28 days, respectively, at 37 °C. Every 48 h, coronal surfaces were rinsed with 200-ppm NaF solution. After the incubation period, all samples were cut with IsoMet 1000 Precision Cutter (Buehler, Lake Bluff, IL, USA) perpendicular to the joint of the material, in the mesio-distal direction. All the tests were performed two times: after 14 and 28 days of incubation, respectively.

2.2. Vickers Microhardness Measurement

The microhardness of samples was determined using the Qness—Q10 M—Microhardness Tester (ATM Qness GmbH, Golling an der Salzach, Austria) using the Vickers method. This method is based on observing the dentine's resistance to plastic deformation. After the incubation period, microhardness was measured on both sides of the specimen. Two values were obtained for each specimen. The measurement was performed using 100 g (HV0.1) for 10 s. Three indents were made on each specimen, and the mean value was calculated. The spacing between the indents was at least three times their diameter. Indents were made in coronal dentine, no further than 200 μm from the material-dentine junction.

2.3. SEM Analysis

SEM analysis was performed on one specimen for each material, after 14 and 28 days, using a Phenom XL Scanning Electron Microscope (Phenom-World BV, Eindhoven, The Netherlands). The acquisition parameters were: 15 KV accelerating voltage, BSD Full detector, 60 Pa low vacuum, and 3840 × 2160 scan size. Before the examination, the samples were sputtered with a 10-mm-thick layer of gold. The surfaces of the sample, as well as the junction between the material and dentine, were observed.

2.4. EDS Analysis

EDS analysis was also conducted on one specimen for each material. The tests were made after 14 and 28 days, this time using the Inca 350 EDS System (Oxford Instruments, High Wycombe, UK). Before the examinations, the samples were gently polished with a soft brush and air-dried.

2.5. Statistical Analysis

The results were analyzed using SPSS 22.0 statistical software (IBM, Armonk, NY, USA). The values of materials and time points were compared using analysis of variance (repeated measures) across different materials and a Scheffe post hoc test to compare groups. The statistical significance level was set to 0.05. Before analysis, the Kolmogorov–Smirnov test was performed on the distributions, and it was established that they do not significantly differ from normal values.

3. Results

The results of microhardness testing in the zone that was covered with nail varnish and was unaffected by the protocols averaged at 66 ± 1.95 (HV0.1) in both testing periods and did not differ significantly among the groups ($p > 0.05$). The mean microhardness values obtained after 14 and 28 days in the zone submitted to the protocols are shown in Table 2.

Table 2. Mean dentine microhardness and SD values (HV0.1) after 14 and 28 days ($p < 0.001$).

Material	After 14 Days	After 28 Days
EQUIA Forte® HT	26.7 ± 1.45	37.74 ± 1.56
Riva Self Cure	19.66 ± 1.02	29.58 ± 1.18
Cention Forte	19.01 ± 1.24	27.93 ± 1.33
Biodentine™	23.35 ± 1.23	29.92 ± 1.02
GC Fuji TRIAGE®	25.94 ± 1.35	33.87 ± 5.57
Control Group	15.57 ± 0.68	15.64 ± 0.82

The mean microhardness values (HV0.1) obtained after 14 days were significantly different between most groups ($p < 0.001$), with several exceptions (Biodentine vs. Cention Forte $p = 0.08$, Biodentine vs. Riva SC $p = 0.997$, Riva Self Cure vs. Cention Forte $p = 0.229$). Similarly, after 28 days, there were statistically significant differences between all groups ($p < 0.001$), except EQUIA Forte HT and GC Fuji TRIAGE ($p = 0.514$) and Cention Forte and Riva Self Cure ($p = 0.687$).

SEM analysis showed uneven patterns, mineral deposits, debris and cracks on the sample's surfaces. Its results are shown in Figures 1 and 2, which represent sample surfaces treated with different examined materials, obtained after 14 and 28 days of incubation.

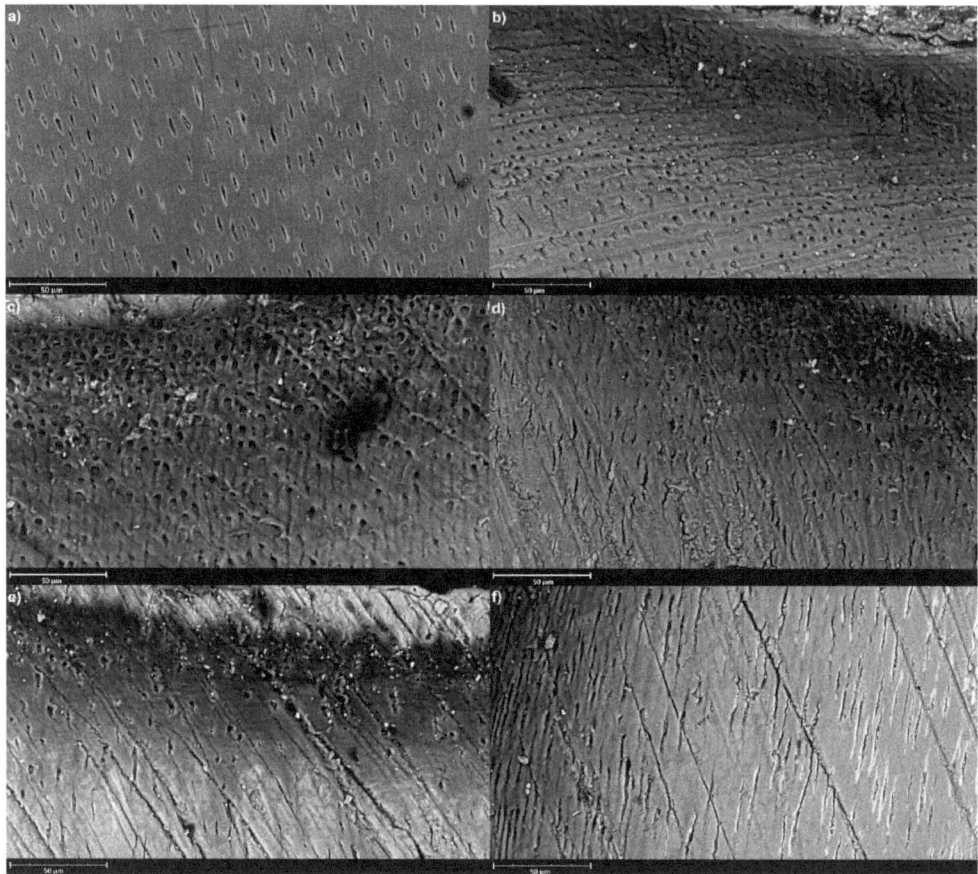

Figure 1. Representative SEM images (accelerating voltage of 15 kV, working distance of 8997 mm, magnification of 2100×, scale bar represents 50 μm) of the sample surfaces after 14 days of incubation; (**a**) Control, (**b**) EQUIA Forte HT, (**c**) GC Fuji TRIAGE, (**d**) Biodentine, (**e**) Riva Self Cure, (**f**) Cention Forte.

Figure 2. Representative SEM images (accelerating voltage of 15 kV, working distance of 8997 mm, magnification of 2100×, scale bar represents 50 μm) of the sample surfaces after 28 days of incubation; (**a**) Control, (**b**) EQUIA Forte HT, (**c**) GC Fuji TRIAGE, (**d**) Biodentine, (**e**) Riva Self Cure, (**f**) Cention Forte.

Figure 3 shows the results of EDS analysis for each group after 14 and 28 days of incubation, with significant element share differences.

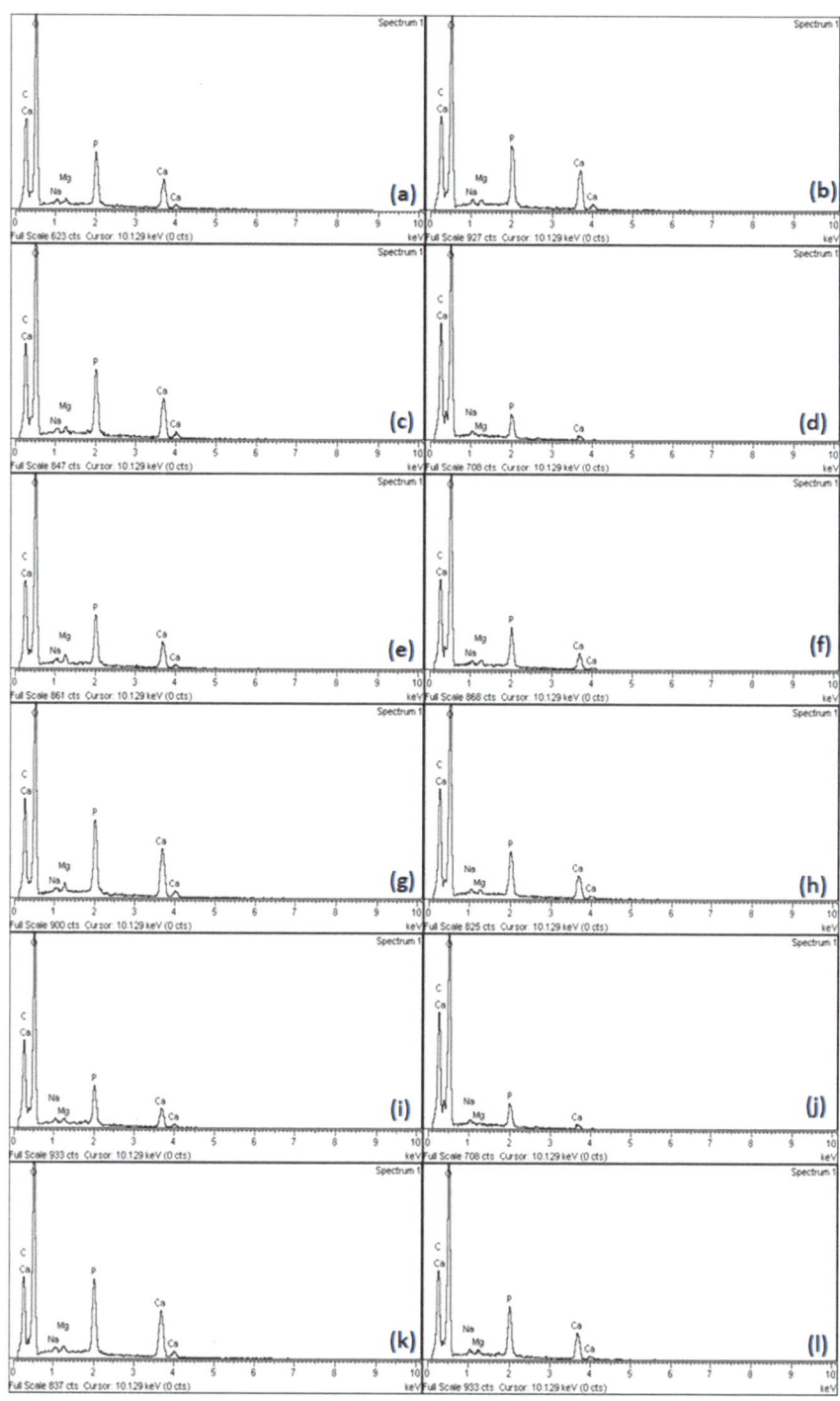

Figure 3. Representative results of EDS elemental analysis after 14- and 28-day incubation period of the samples in (**a**,**b**) the control group, (**c**,**d**) EQUIA Forte HT, (**e**,**f**) GC Fuji TRIAGE, (**g**,**h**) Biodentine, (**i**,**j**) Riva Self Cure and (**k**,**l**) Cention Forte.

4. Discussion

Given the vast number of commercially available products marked as ion-releasing, in this study, we wanted to contrast more recently introduced materials with standard ones. Since the results of the present study showed significant differences among the microhardness values obtained for the tested materials, the first null hypothesis was rejected.

In the present study, the material that performed the best in microhardness testing is EQUIA Forte HT, a glass hybrid, closely followed by GC Fuji TRIAGE, a glass ionomer cement. This was further proven by the EDS analysis, which was used for observing the changes in the mineral content of the samples. EDS analysis gave an insight into the chemical changes that occurred during the demineralization and incubation processes. Following demineralization, a lower share of calcium and phosphate ions was detected, and concentrations were observed, as is expected when mineral loss occurs. On the other hand, in all five groups, increased calcium and phosphate amounts could be observed when analyzing the EDS results after incubation: the highest was in the EQUIA Forte HT group. The possible explanation for this is that due to increased matrix cross-linking, glass hybrids release ions somewhat slower, which eventually results in deeper remineralization of the lesion and higher microhardness values. However, Schwendicke et al. [24] previously did not obtain any mineral gain after the usage of GICs, most likely due to differences in the study design; in the present study, a simpler demineralization protocol was used. Furthermore, the remineralization period in our study was shorter, not allowing precipitated ions to wash off. Several factors possibly contributed to glass hybrid and glass ionomer cements performing well. Glass ionomers are already a well-known remineralizing agent, and our results are similar to others' findings [25,26]. However, in the present study, two glass hybrids performed differently, with much higher values obtained for EQUIA Forte HT. compared to Riva Self Cure. This situation is not unprecedented: Bueno et al. [27] found that not all materials release fluorides and other ions in the same way. These differences are mostly attributed to their composition and the kinetics of their setting reaction [28], as fluoride gets easily trapped within the matrix. The diffusion may vary depending on the material composition and the extent of the reaction, which eventually determines the number of ions released [29]. In contrast, Biodentine also increased microhardness, possibly because it contains particles of a small size of 2811 m^2/g for a quick release of calcium and phosphate ions [30], but its results were inferior to EQUIA Forte HT and GC Fuji TRIAGE, which was furthermore supported by the findings of Al-Abdi et al. [31]. However, it should be considered that this is a material that increases TGF-β1 pulp cell secretion in cell differentiation and mineralization, thus making it more suitable for in vivo studies [32]. The lowest level of mineral gain was observed in the Cention Forte group. Still, it also increased the mineral share to a certain extent. This can be explained by its properties of a low monomer conversion and lower cross-linking density of the resin matrix, which can enhance fluoride diffusion and which have possibly contributed to this result [33,34]. On the other hand, its inferior results to glass hybrids and glass ionomers are possibly due to self-cured materials releasing fluorides at a higher rate than dual-cured materials [35].

SEM/EDS analysis was performed to observe the changes in mineral content. SEM is useful for observing changes in surface structure and appearance, whereas EDS is used for observing the mineral share rise or fall [36]. Taking into consideration that the differences in the appearance of the surface structure and the mineral content of the tested samples after incubation were significant, the second and third null hypotheses were partially rejected. Samples included in this part of the study were very fragile; to reduce the examiner's influence on the outcome, they were only gently polished with a soft brush and air-dried before SEM/EDS, leaving markings made by the saw during the cutting process visible. Furthermore, uneven accumulations of minerals can be seen on all specimens, proving the ion deposition. They penetrate to different extents into the lesion. These differences are most likely due to the unique and complex features of dentine, as its properties vary greatly through the tooth. The assessment of microhardness is widely recognized as a straightforward, dependable, and non-destructive methodology, affording valuable insights into the dynamic processes of demineralization and remineralization [37]. Previous research has proven it to be useful in relation to both dentine and enamel for observing alterations in the mechanical properties of dental tissues, allowing for the characterization of changes occurring at the microstructural level; thus, it was used in the present study [38,39]. In addition, a correlation between the mechanical properties of dental hard tissue and its mineral content was found, making this method useful for determining possible remineralization [34,40–42]. An artificial demineralized lesion was created using a demineralizing solution, which was suggested by Ten Cate et al. [22]. Within the limitations of this study, this was seen as sufficient, since it has been shown that an absolute simulation of oral conditions is hardly achievable, as numerous factors, including the velocity of saliva flow, its buffering capacity, dynamic pH fluctuations in the oral cavity, and behavioral variations, collectively exert substantial influence on an authentic oral environment [43,44]. As expected, the exposure of dentine to the demineralization protocol decreased the microhardness value, which is shown in the control group. As far as the duration of the remineralization course is concerned, various timespans have been used: Bertassoni's protocol lasted several hours [19], while Talwar's protocol lasted 28 days [45]. In the present study, microhardness was measured after 14 and 28 days to obtain results comparable to other studies. All tested materials showed an increase in microhardness in both measuring periods, with higher values obtained after 28 days. This can be explained by the gradual fluoride release, as fluoride ions do not react chemically during the setting reaction, which allows them to diffuse down their concentration gradient [46].

While all the tested materials helped to improve the dentine microhardness, compared to demineralized lesions, the obtained values were still significantly lower than those obtained on the unaffected side. This could be attributed to the fact that pulpal defense and the activity of its cells, such as odontoblasts and fibroblasts is, with the presence of growth factors, crucial for functional mineral deposition [47], which is another limitation of the study. Still, the examined materials, with their ability to deploy ions, can be a valuable clinical asset, due to their buffering effect and anti-microbial effects [48,49]. As the remineralization of dentine is very complex and often results in the heterogenous formation of crystals, great differences between studies are possible, making comparisons challenging. However, such in vitro studies, in which the oral conditions were not utterly simulated, still gave us an idea of the materials' capability for releasing ions, which, together with the pulpal activity, can lead to remineralization. Nevertheless, the results are to be observed while considering the experimental conditions. Therefore, additional research is needed to answer a great number of questions that arise regarding this fast-developing group of materials.

5. Conclusions

Within the limitations of this study, EQUIA Forte HT, a glass hybrid, showed the highest potential for the release of ions into an artificially demineralized lesion, as evidenced by its effect on the microhardness and chemical composition when compared to Riva Self Cure, Cention Forte, Biodentine and GC Fuji TRIAGE. However, all materials tested increased the mineral content of the lesion, making the use of ion-releasing materials promising, along with further research that could include longer exposure periods and different environmental conditions. Finally, conducting comprehensive studies to evaluate the biological responses and clinical outcomes associated with these materials would provide valuable insights to support the effective use of these materials.

Author Contributions: Conceptualization, I.Š. and I.M.; methodology, I.M.; software, F.S.; validation, F.S., H.A. and I.M.; formal analysis, J.Y.; investigation, I.Š. and J.Y.; resources, F.S.; data curation, J.Y.; writing—original draft preparation, I.Š.; writing—review and editing, I.Š. and I.M.; visualization, I.M.; supervision, I.M.; project administration, I.M.; funding acquisition, I.M. All authors have read and agreed to the published version of the manuscript.

Funding: This research was funded by the Croatian Science Foundation project 'Research and Development of New Micro and Nanostructural Bioactive Materials in Dental Medicine', BIODENTMED No. IP-2018-01-1719.

Institutional Review Board Statement: The study was conducted in accordance with the Declaration of Helsinki, and the Ethics Committee of the School of Dental Medicine, University of Zagreb approved the protocol for the current study (05-PA-30-III-12/2021).

Informed Consent Statement: Informed consent was obtained from all subjects involved in the study.

Data Availability Statement: The data presented in this study are available on request from the corresponding author.

Conflicts of Interest: The authors declare no conflict of interest.

References

1. Goldberg, M.; Kulkarni, A.B.; Young, M.; Boskey, A. Dentin: Structure, composition and mineralization. *Front. Biosci.* **2011**, *3*, 711–735. [CrossRef]
2. Tjäderhane, L.; Carrilho, M.R.; Breschi, L.; Tay, F.R.; Pashley, D.H. Overview of dentin structure. *Endod. Top.* **2009**, *20*, 3–29. [CrossRef]
3. Niu, L.-N.; Zhang, W.; Pashley, D.H.; Breschi, L.; Mao, J.; Chen, J.-H.; Tay, F.R. Biomimetic remineralization of dentin. *Dent. Mater.* **2014**, *30*, 77–96. [CrossRef]
4. Bertassoni, L.E.; Habelitz, S.; Kinney, J.H.; Marshall, S.J.; Marshall, G.W., Jr. Biomechanical Perspective on the Remineralization of Dentin. *Caries Res.* **2009**, *43*, 70–77. [CrossRef] [PubMed]
5. Sauro, S.; Osorio, R.; Watson, T.F.; Toledano, M. Influence of phosphoproteins' biomimetic analogs on remineralization of mineral-depleted resin–dentin interfaces created with ion-releasing resin-based systems. *Dent. Mater.* **2015**, *31*, 759–777. [CrossRef] [PubMed]
6. Pires, P.M.; Neves, A.D.A.; Makeeva, I.M.; Schwendicke, F.; Faus-Matoses, V.; Yoshihara, K.; Banerjee, A.; Sauro, S. Contemporary restorative ion-releasing materials: Current status, interfacial properties and operative approaches. *Br. Dent. J.* **2020**, *229*, 450–458. [CrossRef]
7. Christensen, G.J. The advantages of minimally invasive dentistry. *J. Am. Dent. Assoc.* **2005**, *136*, 1563–1565. [CrossRef]
8. Mirsiaghi, F.; Leung, A.; Fine, P.; Blizard, R.; Louca, C. An investigation of general dental practitioners' understanding and perceptions of minimally invasive dentistry. *Br. Dent. J.* **2018**, *225*, 420–424. [CrossRef]
9. Brambilla, E.; Ionescu, A.C. Oral Biofilms and Secondary Caries Formation. In *Oral Biofilms and Modern Dental Materials: Advances toward Bioactivity*; Springer Nature: Basingstoke, UK, 2021; pp. 19–35.
10. Nicholson, J.W. Chemistry of glass-ionomer cements: A review. *Biomaterials* **1998**, *19*, 485–494. [CrossRef]
11. Menezes-Silva, R.; Cabral, R.N.; Pascotto, R.C.; Borges, A.F.S.; Martins, C.C.; de Lima Navarro, M.F.; Sidhu, S.K.; Leal, S.C. Mechanical and optical properties of conventional restorative glass-ionomer cements—A systematic review. *J. Appl. Oral Sci.* **2019**, *27*, e2018357. [CrossRef]
12. Cho, S.Y.; Cheng, A.C. A review of glass ionomer restorations in the primary dentition. *J. Can. Dent. Assoc.* **1999**, *65*, 491–495. [PubMed]
13. Sidhu, S.K.; Nicholson, J.W. A Review of Glass-Ionomer Cements for Clinical Dentistry. *J. Funct. Biomater.* **2016**, *7*, 16. [CrossRef] [PubMed]
14. Hasan, A.M.H.R.; Sidhu, S.K.; Nicholson, J.W. Fluoride release and uptake in enhanced bioactivity glass ionomer cement ("glass carbomer™") compared with conventional and resin-modified glass ionomer cements. *J. Appl. Oral Sci.* **2019**, *27*, e20180230. [CrossRef] [PubMed]
15. Barot, T.; Rawtani, D.; Kulkarni, P. Nanotechnology-based materials as emerging trends for dental applications. *Rev. Adv. Mater. Sci.* **2021**, *60*, 173–189. [CrossRef]
16. Kutuk, Z.; Ozturk, C.; Cakir, F.; Gurgan, S. Mechanical performance of a newly developed glass hybrid restorative in the restoration of large MO Class 2 cavities. *Niger. J. Clin. Pract.* **2019**, *22*, 833–841. [CrossRef]
17. Torabinejad, M.; Hong, C.; McDonald, F.; Pitt Ford, T.R. Physical and chemical properties of a new root-end filling material. *J. Endod.* **1995**, *21*, 349–353. [CrossRef]
18. Shetty, S.; Bhat, R.; Kini, A.; Shetty, P. Microleakage Evaluation of an Alkasite Restorative Material: An In Vitro Dye Penetration Study. *J. Contemp. Dent. Pract.* **2019**, *20*, 1315–1318. [CrossRef]
19. Bertassoni, L.E.; Habelitz, S.; Marshall, S.J.; Marshall, G.W. Mechanical recovery of dentin following remineralization in vitro—An indentation study. *J. Biomech.* **2011**, *44*, 176–181. [CrossRef]
20. Braga, R.R.; Habelitz, S. Current Developments on Enamel and Dentin Remineralization. *Curr. Oral Health Rep.* **2019**, *6*, 257–263. [CrossRef]

21. Bahammam, S.; Nathanson, D.; Fan, Y. Evaluating the Mechanical Properties of Restorative Glass Ionomers Cements. *Int. Dent. J.* **2022**, *72*, 859–865. [CrossRef]
22. Cate, J.M.T.; Damen, J.; Buijs, M. Inhibition of Dentin Demineralization by Fluoride in vitro. *Caries Res.* **1998**, *32*, 141–147. [CrossRef]
23. Tschoppe, P.; Meyer-Lueckel, H.; Kielbassa, A. Effect of carboxymethylcellulose-based saliva substitutes on predemineralised dentin evaluated by microradiography. *Arch. Oral Biol.* **2008**, *53*, 250–256. [CrossRef] [PubMed]
24. Schwendicke, F.; Al-Abdi, A.; Moscardó, A.P.; Cascales, A.F.; Sauro, S. Remineralization effects of conventional and experimental ion-releasing materials in chemically or bacterially-induced dentin caries lesions. *Dent. Mater.* **2019**, *35*, 772–779. [CrossRef] [PubMed]
25. Ngo, H.C.; Mount, G.; Mc Intyre, J.; Tuisuva, J.; Von Doussa, R. Chemical exchange between glass-ionomer restorations and residual carious dentine in permanent molars: An in vivo study. *J. Dent.* **2006**, *34*, 608–613. [CrossRef] [PubMed]
26. Ten Cate, J.M.; Buus, J.M.; Damen, J.M. The effects of GIC restorations on enamel and dentin demineralization and remineralization. *Adv. Dent. Res.* **1995**, *9*, 384–388. [CrossRef]
27. Bueno, L.S.; Silva, R.M.; Magalhães, A.P.R.; Navarro, M.F.L.; Pascotto, R.C.; Buzalaf, M.A.; Nicholson, J.W.; Sidhu, S.K.; Borges, A.F.S. Positive correlation between fluoride release and acid erosion of restorative glass-ionomer cements. *Dent. Mater.* **2019**, *35*, 135–143. [CrossRef] [PubMed]
28. De Moor, R.J.; Verbeeck, R.M.; De Maeyer, E.A. Fluoride release profiles of restorative glass ionomer formulations. *Dent. Mater.* **1996**, *12*, 88–95. [CrossRef]
29. Griffin, S.; Hill, R. Influence of glass composition on the properties of glass polyalkenoate cements. Part IV: Influence of fluorine content. *Biomaterials* **2000**, *21*, 693–698. [CrossRef]
30. Luo, Z.; Li, D.; Kohli, M.R.; Yu, Q.; Kim, S.; He, W.-X. Effect of Biodentine™ on the proliferation, migration and adhesion of human dental pulp stem cells. *J. Dent.* **2014**, *42*, 490–497. [CrossRef]
31. Al-Abdi, A.; Paris, S.; Schwendicke, F. Glass hybrid, but not calcium hydroxide, remineralized artificial residual caries lesions in vitro. *Clin. Oral Investig.* **2016**, *21*, 389–396. [CrossRef]
32. Laurent, P.; Camps, J.; About, I. Biodentine™ induces TGF-β1 release from human pulp cells and early dental pulp mineralization. *Int. Endod. J.* **2011**, *45*, 439–448. [CrossRef]
33. Panpisut, P.; Toneluck, A. Monomer conversion, dimensional stability, biaxial flexural strength, and fluoride release of resin-based restorative material containing alkaline fillers. *Dent. Mater. J.* **2020**, *39*, 608–615. [CrossRef]
34. Theerarath, T.; Sriarj, W. An alkasite restorative material effectively remineralized artificial interproximal enamel caries in vitro. *Clin. Oral Investig.* **2022**, *26*, 4437–4445. [CrossRef]
35. Donly, K.J.; Liu, J.A. Dentin and enamel demineralization inhibition at restoration margins of Vitremer, Z 100 and Cention N. *Am. J. Dent.* **2018**, *31*, 166–168.
36. Shaik, Z.A.; Rambabu, T.; Sajjan, G.; Varma, M.; Satish, K.; Raju, V.B.; Ganguru, S.; Ventrapati, N. Quantitative Analysis of Remineralization of Artificial Carious Lesions with Commercially Available Newer Remineralizing Agents Using SEM-EDX- In Vitro Study". *J. Clin. Diagn. Res.* **2017**, *11*, ZC20–ZC23. [CrossRef]
37. Mousavinasab, S.M.; Meyers, I. Fluoride Release by Glass Ionomer Cements, Compomer and Giomer. *Dent. Res. J.* **2009**, *6*, 75–81.
38. Almahdy, A.; Downey, F.; Sauro, S.; Cook, R.; Sherriff, M.; Richards, D.; Watson, T.; Banerjee, A.; Festy, F. Microbiochemical Analysis of Carious Dentine Using Raman and Fluorescence Spectroscopy. *Caries Res.* **2012**, *46*, 432–440. [CrossRef] [PubMed]
39. Salinovic, I.; Schauperl, Z.; Marcius, M.; Miletic, I. The Effects of Three Remineralizing Agents on the Microhardness and Chemical Composition of Demineralized Enamel. *Materials* **2021**, *14*, 6051. [CrossRef] [PubMed]
40. Angker, L.; Nockolds, C.; Swain, M.V.; Kilpatrick, N. Correlating the mechanical properties to the mineral content of carious dentine—A comparative study using an ultra-micro indentation system (UMIS) and SEM-BSE signals. *Arch. Oral Biol.* **2004**, *49*, 369–378. [CrossRef] [PubMed]
41. Shetty, S.; Hegde, M.; Bopanna, T. Enamel remineralization assessment after treatment with three different remineralizing agents using surface microhardness: An in vitro study. *J. Conserv. Dent.* **2014**, *17*, 49–52. [CrossRef]
42. Joshi, C.; Gohil, U.; Parekh, V.; Joshi, S. Comparative evaluation of the remineralizing potential of commercially available agents on artificially demineralized human enamel: An In vitro study. *Contemp. Clin. Dent.* **2019**, *10*, 605–613. [CrossRef]
43. Damato, F.; Strang, R.; Stephen, K. Effect of Fluoride Concentration on Remineralization of Carious Enamel an in vitro pH-Cycling Study. *Caries Res.* **1990**, *24*, 174–180. [CrossRef]
44. Marsh, P.D. Microbiologic aspects of dental plaque and dental caries. *Dent. Clin. N. Am.* **1999**, *43*, 599–614. [CrossRef] [PubMed]
45. Talwar, M.; Borzabadi-Farahani, A.; Lynch, E.; Borsboom, P.; Ruben, J. Remineralization of Demineralized Enamel and Dentine Using 3 Dentifrices—An InVitro Study. *Dent. J.* **2019**, *7*, 91. [CrossRef] [PubMed]
46. Rajić, V.B.; Miletić, I.; Gurgan, S.; Peroš, K.; Verzak, Ž.; Malčić, A.I. Fluoride Release from Glass Ionomer with Nano Filled Coat and Varnish. *Acta Stomatol. Croat.* **2018**, *52*, 307–313. [CrossRef] [PubMed]
47. Galler, K.M.; Weber, M.; Korkmaz, Y.; Widbiller, M.; Feuerer, M. Inflammatory Response Mechanisms of the Dentine–Pulp Complex and the Periapical Tissues. *Int. J. Mol. Sci.* **2021**, *22*, 1480. [CrossRef] [PubMed]

48. Nyvad, B.; Machiulskiene, V.; Baelum, V. Reliability of a New Caries Diagnostic System Differentiating between Active and Inactive Caries Lesions. *Caries Res.* **1999**, *33*, 252–260. [CrossRef]
49. Slimani, A.; Sauro, S.; Hernández, P.G.; Gurgan, S.; Turkun, L.S.; Miletic, I.; Banerjee, A.; Tassery, H. Commercially Available Ion-Releasing Dental Materials and Cavitated Carious Lesions: Clinical Treatment Options. *Materials* **2021**, *14*, 6272. [CrossRef]

Disclaimer/Publisher's Note: The statements, opinions and data contained in all publications are solely those of the individual author(s) and contributor(s) and not of MDPI and/or the editor(s). MDPI and/or the editor(s) disclaim responsibility for any injury to people or property resulting from any ideas, methods, instructions or products referred to in the content.

Article

Comparison of Orthodontic Tooth Movement of Regenerated Bone Induced by Carbonated Hydroxyapatite or Deproteinized Bovine Bone Mineral in Beagle Dogs

Takaharu Abe [1], Ryo Kunimatsu [2,*] and Kotaro Tanimoto [2]

1. Department of Orthodontics, Division of Oral Health and Development, Hiroshima University Hospital, Hiroshima 734-8553, Japan; takabe@hiroshima-u.ac.jp
2. Department of Orthodontics and Craniofacial Developmental Biology, Hiroshima University Graduate School of Biomedical & Health Sciences, Hiroshima 734-8553, Japan; tkotaro@hiroshima-u.ac.jp
* Correspondence: ryoukunimatu@hiroshima-u.ac.jp; Tel.: +81-82-257-5686; Fax: +81-82-257-5687

Abstract: Orthodontic treatments often involve tooth movement to improve dental alignment. In this study, we aimed to compare tooth movement in regenerated bone induced by two different bone fillers, carbonated hydroxyapatite (CAP) and deproteinized bovine bone mineral (DBBM). Four beagle dogs were used in this comparative study. The first, second, and fourth lower mandibular premolars (P1, P2, and P4) on both sides of the mouth were extracted, and CAP was implanted into the extraction site on the left side and DBBM into the right side. Following regenerative bone healing, orthodontic devices were attached to perform orthodontic tooth movement of the lower third mandibular premolar (P3) on both sides. X-ray examination, intraoral scan, and histological analysis were performed. The Mann–Whitney U test was used for statistical analysis, and $p < 0.05$ was considered significant. Bone regeneration and orthodontic tooth movement were observed in the CAP and DBBM groups. Histologically, normal periodontal tissue remodeling was observed on the compression and tension sides of CAP and DBBM. No statistical difference was observed in the number of osteoclasts around the periodontal ligament and the root resorption area. Orthodontic tooth movement of regenerated bone induced by CAP and DBBM was therefore achieved.

Keywords: bone regeneration; carbonated hydroxyapatite; deproteinized bovine bone mineral; orthodontic tooth movement

1. Introduction

Various bone filling materials, such as collagen gel [1], deproteinized bovine bone mineral (DBBM) [2], hydroxyapatite (HA) [3], and β-tricalcium phosphate (β-TCP) [4], have been used for alveolar bone regeneration and dental treatment.

For orthodontic purposes, bone filler is typically transformed into physiological bone after it is grafted into the body, serving its intended metabolic purpose. In contrast to bone augmentation in implant therapy, a bone filler placed into a tooth socket (followed by orthodontic treatment) must be bioabsorbable to allow tooth movement.

Of the bone filler candidates, collagen exhibits disadvantages such as low physical strength and difficulty maintaining its shape within the oral cavity. Similarly, HA does not allow orthodontic tooth movement at the implantation site because it cannot be replaced by bone in the long term [5]. Furthermore, β-TCP is hydrolyzed and dissolved in body fluids, resulting in bone regeneration while maintaining its original shape and density [6].

Jensen et al. [2] reported that DBBM implantation can effectively regenerate bones. DBBM was approved in Japan in 1999 and is widely used as a bone augmentation material for implant placement [7]. Studies indicate that DBBM will eventually be replaced by natural bone over time [8,9]. Araújo et al. [10] reported that DBBM did not interfere with tooth movement if it was implanted in the tooth extraction site and orthodontic force was applied.

Carbonated hydroxyapatite (CAP) is another artificial biomaterial that was approved for clinical use in Japan in 2017 [11]. CAP contains 6–9% carbonate in its apatite structure and exhibits high osteoconductivity and remodeling ability [11]. Mano et al. [12] observed alveolar ridge continuity after CAP implantation. Fukuda et al. [13] demonstrated that CAP was effective in maintaining the trabecular bone after tooth extraction. Therefore, owing to these favorable properties, CAP is frequently used in dental treatment. CAP is currently being applied clinically as a bone filler for bone augmentation and regeneration of periodontal tissue in implant-bearing regions [14,15].

It has been previously reported that regenerated bone induced by CAP and mesenchymal stem cells (MSCs) can produce orthodontic tooth movement [16]. However, no studies have investigated tooth movement in regenerated bone using CAP without MSCs and compared it with other bone filler materials. Therefore, in this study, we aimed to evaluate the effects of DBBM and CAP as bone filling materials on bone regeneration and tooth movement.

2. Materials and Methods

This study was approved by the relevant ethics review board of Hamley Co., Ltd. (Ibaraki, Japan) Animal Ethics Committee [Study No. 20-H022].

2.1. Scaffold Implantation

Four dogs (TOYO beagles, 12 months old; Kitayama Labes, Nagano, Japan) were anesthetized via intramuscular injection with 0.4 mL/kg of ketamine hydrochloride (Ketalal 500 mg, Daiichi Sankyo Propharma, Tokyo, Japan) and xylazine (Theratal 2% injection, Bayer Yakuhin, Osaka, Japan) in a 1:1 mixed solution. The same solution was also administered to maintain anesthesia.

The operation area was sterilized with an Isodine solution for animals (Mundipharma Co., Ltd., Tokyo, Japan; 20 mg of Japanese Pharmacopoeia povidone-iodine in 1 mL), and xylocaine was used to induce local anesthesia. The first, second, and fourth premolars (P1, P2, and P4) of the mandible were then extracted. CAP obtained from Cytrans Granules® (GC, Tokyo, Japan) and DBBM obtained from Bio-Oss® (Geistlich Pharma AG, Wolhusen, Switzerland) were used for the experiments. The bone filling material was implanted in the extraction sites and the gingiva was sutured to close the wound. DBBM S (0.25–1.0 mm in size) was implanted on the right side, and CAP S (0.3–0.6 mm) and M (0.6–1.0 mm) mixed at a 1:1 ratio were implanted on the left side. Healing progress was checked monthly for 3 months after Isodine disinfection.

2.2. Attaching Orthodontic Appliances and Moving Teeth

Healing of the extraction site was confirmed 6 weeks after transplantation, and an impression was taken using silicone material (Examixfine Regular type; GC) and a tray. Three months after implantation, an orthodontic appliance manufactured by Wada Seimitsu Laboratory using a casting method was attached. We used a stainless steel wire measuring 0.019×0.025 inches for the appliance, and for the crown on P3, we used 0.022-inch slotted tubing. Subsequently, the closed coil spring (TOMY International, Tokyo, Japan) was adjusted to 100 g, and experimental tooth movement was performed. The date of device attachment was defined as day 0, and observations were conducted every 2 months.

2.3. Intraoral and Radiographic Photographs

After tooth extraction and transplantation, radiographs were taken every month for 3 months to confirm healing and bone regeneration. Dental radiographs and intraoral photographs were taken at the start of the orthodontic movement and every 2 months thereafter. Intraoral photographs were captured in the same direction.

2.4. Evaluation of Tooth Movement by an Intraoral Scanner (IOS)

Tooth movement was recorded at the start of the movement and every 2 months after using an intraoral scanner (IOS) (Adva IOS 100, GC, Tokyo, Japan), and Viewbox 4 (version 4.1.0.12) (dHAL Software, Kifissia, Greece) was used to assess tooth movement. The IOS was focused on the canine (C) and first molar (M1): the Z-axis was the center of P3 when the device was attached; the Y-axis was the direction of the wire; and the X-axis followed the occlusal plane from the intersection of the Y and Z axes. Using these planes, the movement distance and inclination angle of the teeth were observed and recorded. Figure 1 depicts the three-dimensional (3D) superimposition of the IOS images before and after tooth movement using Viewbox 4. It should be noted that the evaluation of tooth movement was excluded from the results of this study because a large tilt of the crown was observed in one dog due to a wire fracture.

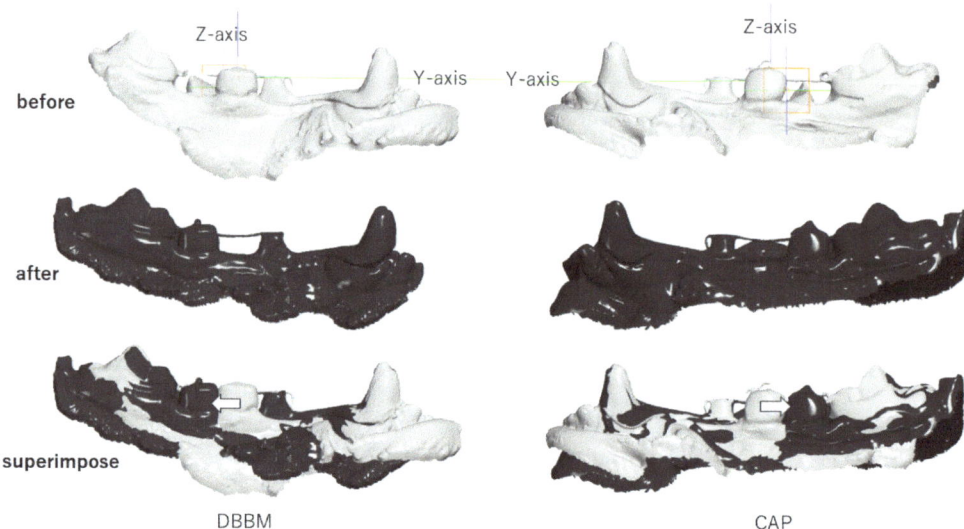

Figure 1. The STL data were superimposed, mainly for the canine (C) and first molar (M1). The Z-axis was the center of P3 with the device attached. The Y-axis was the direction of the wire. The blue line indicates the Z-axis, the green line indicates the Y-axis, and the orange square line indicates the X-axis direction. The white arrow indicates the direction of movement of P3.

2.5. Histological Evaluation of Regenerated Bone

At the end of the movement, an excessive amount of thiamylal sodium (Isososol, Nichi-Iko, Tokyo, Japan) was intravenously administered (50 mg/kg or more) to euthanize the dogs. Excess bone, teeth, and soft tissue were removed after harvesting. Tissues were fixed with 4% Paraformaldehyde (PFA), decalcified with ethylenediaminetetraacetic acid (EDTA) for 1 month. Sections were stained with hematoxylin and eosin (H and E), Masson's trichrome (MT), and tartrate-resistant acid phosphatase (TRAP) after being sliced into 5-μm thick slices. The sagittal section was cut between the P3 and the first posterior molar. The number of TRAP-positive cells per unit area of alveolar bone on the compression side of the moved tooth was counted. For H and E staining, the tooth root resorption area on the compression side of the moved tooth was measured using ImageJ software https://imagej.net/ij/download.html (accessed 22 December 2023) (National Institute of Health, Bethesda, MD, USA).

2.6. Statistical Analysis

Results are presented as the mean ± standard deviation (SD). The Mann–Whitney U test was used for statistical analysis and $p < 0.05$ was considered significant.

3. Results

3.1. Intraoral and Radiographic Evaluation

Intraoral photographs and dental radiographs obtained during tooth extraction and bone filler grafting are shown in Figure 2a,b. One month after implantation, the implanted bone filler was replaced with bone in both groups; trabecular bone formation was observed 2 months after implantation. Intraoral photographs and dental X-ray images captured during the orthodontic tooth movement every 2 months are shown in Figure 2c.

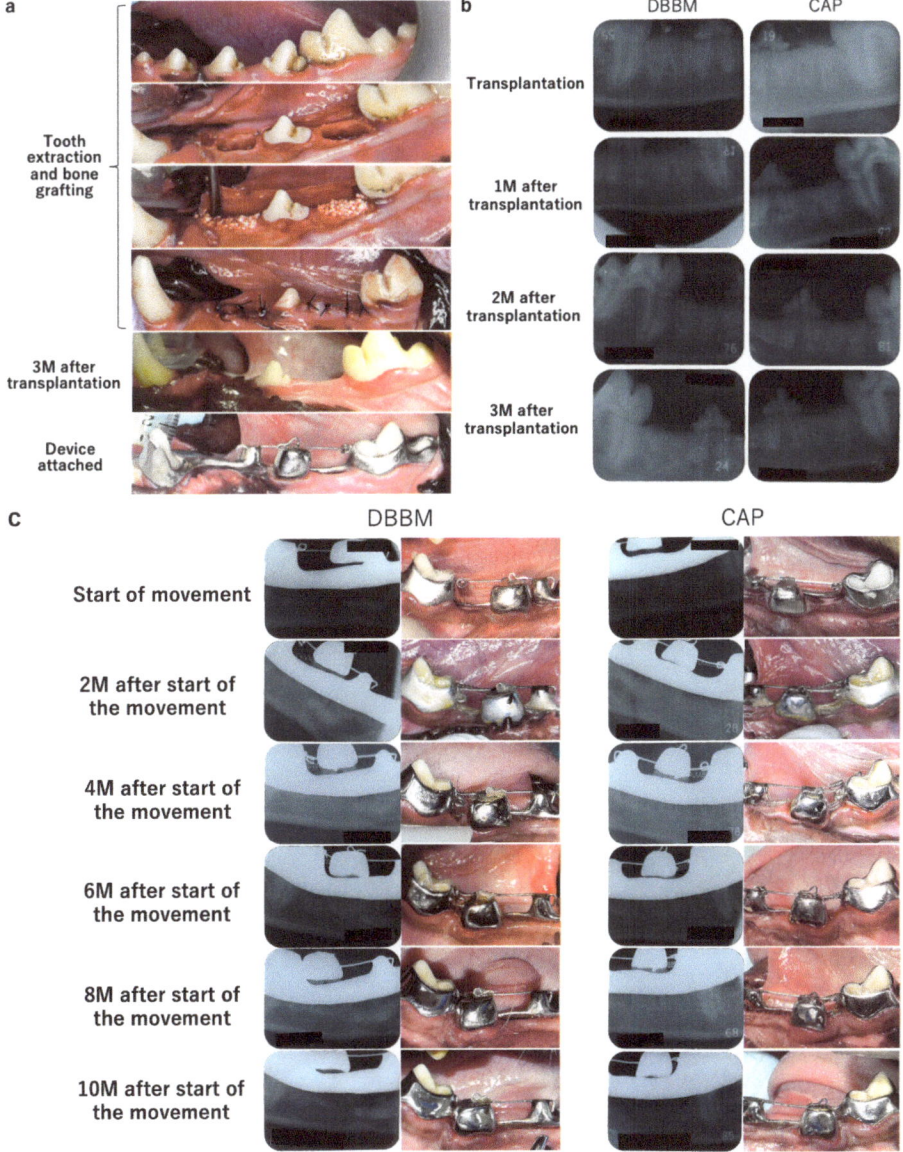

Figure 2. Intraoral photographs and dental radiographs obtained during tooth extraction and bone filler grafting. (**a**) Intraoral photograph. Tooth extraction and carrier implantation were performed.

The state of healing 3 months after transplantation and a photograph of the attached device are shown. (**b**) Dental radiographs at the time of transplantation and 1-, 2-, and 3 months post-transplantation. DBBM was implanted on the right side, and CAP was implanted on the left side. Two months after the transplantation, trabecular formation was observed. (**c**) Intraoral and dental X-ray photographs every 2 months after the movement started.

3.2. 3D Tooth Movement Evaluation

The evaluation of the tooth movement distance and inclination are shown in Figure 3a,b. There were no marked differences between the DBBM and CAP groups in terms of tooth movement distance. At the total distal inclination degree, the distal tooth inclination in the DBBM group tended to be higher than that in the CAP group.

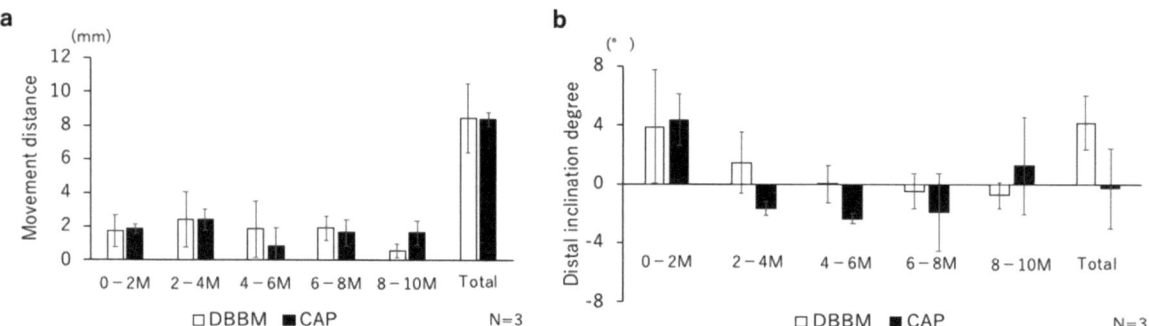

Figure 3. (**a**) Migration distance of P3 along the Y-axis during centrifugation. No significant differences were observed for either schedule. (**b**) Rotation angle toward centrifugation around the X-axis. DBBM tended to have a larger centrifugal slope than CAP but no significant differences were observed. The total movement and inclination indicate the changes between 0 M and 10 M.

3.3. Histological Evaluation

H&E and MT staining images on the compression and traction sides are shown in Figure 4a,b. On the compressed side, the periodontal ligament fibers were disturbed; resorption was observed in part of the alveolar bone, cementum, and dentin. On the tension side, periodontal ligament fibers were elongated, blood vessel permeability was enhanced, and bone formation was observed perpendicular to the long axis of the tooth root. Following MT staining, the periodontal ligament was stained blue, and layers of bone formation were observed around the bone filler. Newly formed bone was stained blue on the traction side.

Most of TRAP-positive cells were observed around the alveolar bone on the compression side of the moved tooth (Figure 5a). The number of TRAP-positive cells observed around the compressed periodontal ligament in the DBBM and CAP groups was 39.0 ± 26.6 and 53.0 ± 17.7, respectively. The root resorption area was 0.23 ± 0.21 mm^2 in DBBM and 0.62 ± 0.22 mm^2 in CAP. CAP tended to have a larger area resorption, but there was no statistical difference between the two groups (Figure 5b).

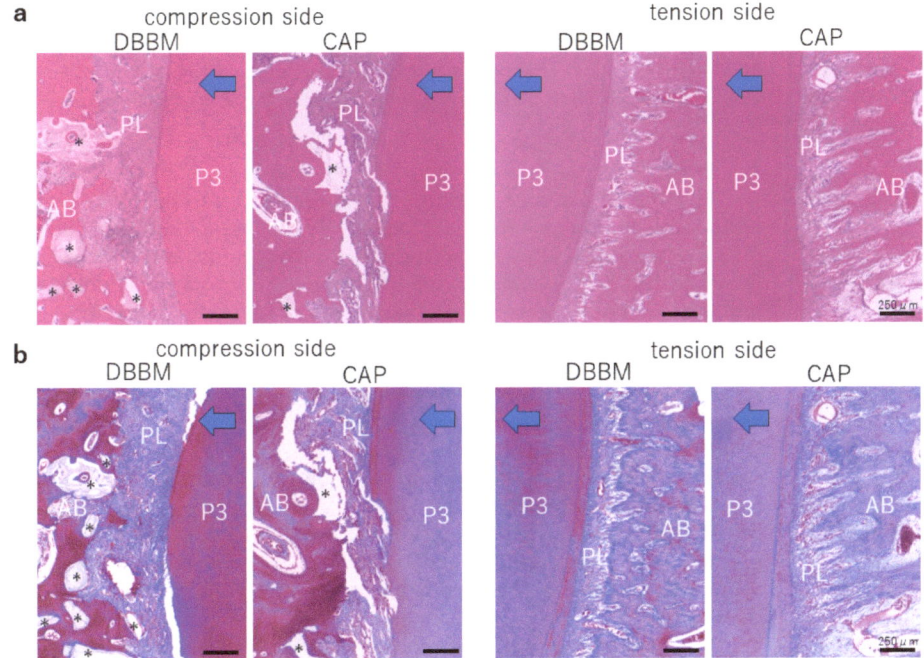

Figure 4. (**a**) H and E staining. On the compressed side, resorption was observed in part of the alveolar bone, cementum, and dentin. On the tension side, bone formation was observed. (**b**) MT staining. M1 teeth are shown. The bone is formed in layers around the bone filler. The newly formed bone is stained blue on the traction side. The blue arrow indicates the direction of tooth traction. * indicates bone filler, P3 indicates the third premolar AB indicates alveolar bone, and PL indicates periodontal ligament; scale bar = 250 μm.

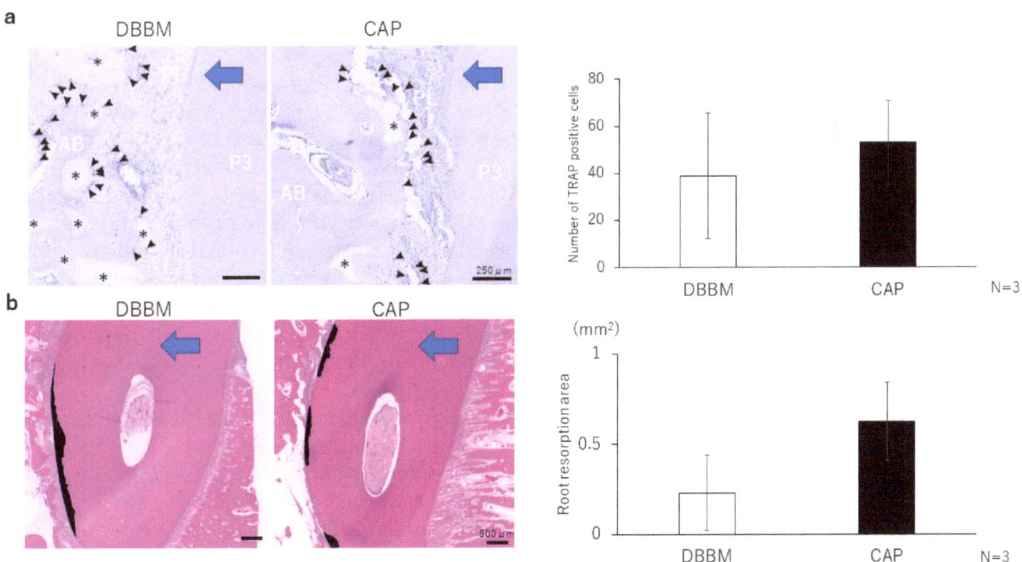

Figure 5. (**a**) TRAP staining. P3 teeth are shown. The arrowheads indicate TRAP-positive osteoclasts. Osteoclasts were observed around the alveolar bone on the pressing side and around the bone filler.

No marked difference was observed in TRAP-positive cells. * indicates bone filler, P3 indicates the third premolar AB indicates alveolar bone, and PL indicates periodontal ligament; Scale bar = 250 µm. (**b**) Black areas indicate resorbed roots. No significant difference was observed in the root resorption area. Scale bar = 500 µm. The blue arrow indicates the direction of tooth traction.

4. Discussion

In this study, a bone substitute was implanted into the tooth extraction site, and orthodontic tooth movement was performed. Filling a tooth extraction site with bone filler has been clinically applied for socket preservation, and its effectiveness has been demonstrated [17]. Numerous reports have documented the application of bone filler implants for bone regeneration [18]. In this study, orthodontic tooth movement was performed on regenerated bone.

It has been suggested that the carbonate content of bone filler is one of the factors contributing to its enhanced osteoconductivity [19]. Furthermore, CAP is stable in vivo and is rapidly absorbed in the weakly acidic environment (pH 5.3) of osteoclast Howship's lacunae. The absorption of carriers by osteoclasts sprouts calcium and carbonate ions, and the increase in extracellular calcium ion concentrations promotes mineralization of the matrix by osteoblasts [20]. Sato et al. [21] conducted a study in which CAP, β-TCP, HA, and DBBM were implanted into a three-walled bone defect and reported that CAP implantation resulted in the rapid recruitment of osteoclasts and endothelial cells. The rapid induction of bone regeneration by the early migration of osteoclasts is an important requirement for graft carriers.

Two months after transplantation, resorption of bone fillers and formation of trabecular bone due to bone regeneration were observed in both the CAP and DBBM groups via X-ray examination. Regarding tooth movement and root resorption, only a few granules were retained, and a complete replacement with bone was expected. However, resorption is caused by osteoclasts generated by tooth movement. Ishikawa et al. [11] reported that CAP was more susceptible to osteoclast resorption than DBBM, but the difference was not significant.

When tooth movement was evaluated using a 3D model, no differences were observed in the distance of tooth movement between the DBBM and CAP groups. Orthodontic tooth movement is achieved by bone metabolism of the alveolar bone in the periodontium, and usually achieved clinically by inclining the crown toward the tooth movement. Subsequently, bone resorption of the alveolar bone at the root of the teeth occurs, leading to migration of the root. However, the mode of tooth movement may vary depending on the environment in the body and the placement status of the bracket. Specifically, depending on bone turnover of the alveolar bone, there may be cases in which the translation is beautiful angle without an inclination occurring, or the roots may be moved first, depending on the angle of the bracket. In this study, we used an SS wire with highly rigid properties, which is compatible with individual teeth, and an engineered device for experimental tooth movement. Angular wires with a size of 0.019×0.025 inches, which is large in clinical terms, and a real multibracket device were used to ensure as little inclination as possible. However, it is known that even hard wires can bend [22]. In both groups, the angle of tilt was up to four degrees, and the incidence of tilt was minimal, which is within the clinically possible extent of tooth movement.

Both bone fillers exhibited high biocompatibility, and a prolonged time period was required for complete bone replacement. Regarding the remaining bone filler, it is possible that osteoclasts on the pressure side absorbed the bone filler during orthodontic tooth movement and induced bone remodeling. However, by implanting bone filler and integrating orthodontic tooth movement, bone formation occurred while maintaining the alveolar bone level. Therefore, it is important to use correctly remodeled bone filler without impeding tooth movement or causing excessive root resorption by osteoclasts. It is known that root resorption occurs in more than 90% of cases as teeth move [23]. There are limited studies on the histological evaluation of the metabolism of bone grafting materials when teeth are moved after implantation. In this study, the bone fillers remained on the compression side and were remodeled by osteoclasts simultaneously with the alveolar bone. Araújo et al. [10] reported that the decomposition rate of bone filler increased on the compression side. Ma et al. [24] similarly reported that orthodontic force promoted the decomposition of bone substitute material, and that it was most effective after 4 weeks. Alalola et al. [25] reported that residual bone grafting materials restrict tooth movement and cause tooth root resorption as a side effect. However, most of the studies have focused on DBBM and β-TCP, and there are few reports on tooth movement using CAP. In the present study, a trend towards a larger root resorption area was observed in CAP compared to DBBM. As reported by Nan et al. [26], increased bone density at the graft site may have caused root resorption. However, as stated in the limitations, our results were not significantly different, and due to the small sample size, further research on root resorption is needed with respect to CAP transplantation.

To improve bone formation, the application of MSCs with CAP can be used to enhance bone regeneration [27]. Suzuki et al. [28] reported that the use of immature osteoblasts in combination with a 3D scaffold resulted in good bone regeneration. Furthermore, when combined with FGF-2, CAP has shown promising results for treating periodontal tissues [15]. As previously stated [1,29,30] (Putranti et al., 2022; Hiraki et al., 2020b; Nakajima et al., 2018), in the future, greater precision of bone regeneration is expected, especially with the use of stem cells and their supernatant as a transplant carrier. In this study, we demonstrated that bone regeneration and tooth movement is possible by implanting CAP and DBBM into a tooth extraction site. An inherent limitation of this study's design was the small sample size. We investigated dogs in this study to evaluate the scaffold through experimental tooth movement in larger animals that are more similar to humans. In accordance with Russell and Burch's 3R principle, we used four dogs ($n = 4$). Moreover, major damage to the device occurred, so that one animal dropped out of the study. We performed our experiments under conditions as identical as possible, but there were individual differences in the actual movement of teeth in vivo due to metabolic differences. Therefore, the explanation of our study may have been confounded by the small sample size. Although CAP and DBBM are approved bone grafting materials for implantation in the human body [2,11], insufficient knowledge has been obtained on orthodontic tooth movement. The demonstration that tooth movement can be achieved in CAP may open further prospects for bone augmentation in patients with cleft lip/palate and bone defects. Future research on tooth movement models in humans should be developed through specific clinical studies.

5. Conclusions

Carbonated hydroxyapatite (CAP) and deproteinized bovine bone mineral (DBBM) were implanted into empty tooth sockets, and bone regeneration was observed; orthodontic tooth movement was initiated and observed. No significant difference was observed between the movement speed of the teeth and the number of osteoclasts induced on the compression side and root resorption area. These findings demonstrate that tooth movement is possible following bone regeneration using CAP and DBBM.

Author Contributions: Conceptualization, R.K. and K.T.; methodology, R.K. and K.T.; validation, T.A. and K.T.; formal analysis, T.A.; investigation, T.A., R.K. and K.T.; resources, T.A. and R.K.; data curation, T.A., R.K. and K.T.; writing—original draft preparation, T.A.; writing—review and editing, R.K. and K.T.; visualization, T.A. and R.K.; supervision, K.T.; project administration, R.K. and K.T.; funding acquisition, R.K. and K.T. All authors have read and agreed to the published version of the manuscript.

Funding: This research was funded by GC Corporation (Tokyo, Japan). This work was supported by Grants-in-Aid for Scientific Research (grant number 23K09415) from the Japan Society for the Promotion of Science.

Institutional Review Board Statement: The animal study protocol was approved by the Animal Ethics Committee of Hamley Co., Ltd. (Ibaraki, Japan) (protocol code No. 20-H022; 13 February 2020).

Data Availability Statement: The datasets used and/or analyzed during the current study are available from the corresponding author on reasonable request.

Acknowledgments: This study was conducted in collaboration with GC Co. Hijiri Masuda, Yoko Ishihara, Chiaki Oizumi, Makoto Hirohara, and Yuhiro Sakai assisted with data curation, funding acquisition, investigation, methodology, project administration, and resources. We are deeply grateful to them. This study was conducted at the HAMRI Co., Ltd. Tsukuba Research Institute Centre. We would like to thank Azusa Seki, and the staff from the Tsukuba Research Institute Centre.

Conflicts of Interest: The authors declare no conflict of interests.

References

1. Hiraki, T.; Kunimatsu, R.; Nakajima, K.; Abe, T.; Yamada, S.; Rikitake, K.; Tanimoto, K. Stem cell-derived conditioned media from human exfoliated deciduous teeth promote bone regeneration. *Oral Dis.* **2020**, *26*, 381–390. [CrossRef]
2. Jensen, T.; Schou, S.; Stavropoulos, A.; Terheyden, H.; Holmstrup, P. Maxillary sinus floor augmentation with Bio-Oss or Bio-Oss mixed with autogenous bone as graft: A systematic review. *Clin. Oral Implant. Res.* **2012**, *23*, 263–273. [CrossRef]
3. Cosso, M.G.; De Brito, R.B.; Piattelli, A.; Shibli, J.A.; Zenóbio, E.G. Volumetric dimensional changes of autogenous bone and the mixture of hydroxyapatite and autogenous bone graft in humans maxillary sinus augmentation. A multislice tomographic study. *Clin. Oral Implant. Res.* **2014**, *25*, 1251–1256. [CrossRef]
4. Guillaume, B. Filling bone defects with β-TCP in maxillofacial surgery: A review. *Morphologie.* **2017**, *101*, 113–119. [CrossRef]
5. Sugimoto, A.; Ohno, K.; Michi, K.; Kanegae, H.; Aigase, J.; Tachikawa, T. Effect of calcium phosphate ceramic particle insertion on tooth eruption. *Oral Surg. Oral Med. Oral Pathol.* **1993**, *76*, 141–148. [CrossRef]
6. Wiltfang, J.; Merten, H.A.; Schlegel, K.A.; Schultze-Mosgau, S.; Kloss, F.R.; Rupprecht, S.; Kessler, P. Degradation characteristics of α and β tri-calcium-phosphate (TCP) in minipigs. *J. Biomed. Mater. Res.* **2002**, *63*, 115–121. [CrossRef]
7. Moreira, C.; Silva, J.R.; Samico, P.; Nishioka, G.N.d.M.; Nishioka, R.S. Application of bio-oss in tissue regenerative treatment prior to implant installation: Literature review. *Braz. Dent. Sci.* **2019**, *22*, 147–154. [CrossRef]
8. Berglundh, T.; Lindhe, J. Healing around implants placed in bone defects treated with Bio-Oss: An experimental study in the dog. *Clin. Oral Implant. Res.* **1997**, *8*, 117–124. [CrossRef]
9. Vasilyev, A.V.; Kuznetsova, V.S.; Bukharova, T.B.; Grigoriev, T.E.; Zagoskin, Y.D.; Galitsina, E.V.; Fatkhudinova, N.L.; Babichenko, I.I.; Chvalun, S.N.; Goldstein, D.V.; et al. Osteoinductive potential of highly porous polylactide granules and Bio-Oss impregnated with low doses of BMP-2. *IOP Conf. Ser. Earth Environ. Sci.* **2020**, *421*, 052035. [CrossRef]
10. Araújo, M.G.; Carmagnola, D.; Berglundh, T.; Thilander, B.; Lindhe, J. Orthodontic movement in bone defects augmented with Bio-Oss. An experimental study in dogs. *J. Clin. Periodontol.* **2001**, *28*, 73–80. [CrossRef]
11. Ishikawa, K.; Hayashi, K. Carbonate apatite artificial bone. *Sci. Technol. Adv. Mater.* **2021**, *22*, 683–694. [CrossRef]
12. Mano, T.; Akita, K.; Fukuda, N.; Kamada, K.; Kurio, N.; Ishikawa, K.; Miyamoto, Y. Histological comparison of three apatitic bone substitutes with different carbonate contents in alveolar bone defects in a beagle mandible with simultaneous implant installation. *J. Biomed. Mater. Res. B Appl. Biomater.* **2020**, *108*, 1450–1459. [CrossRef]
13. Fukuda, N.; Ishikawa, K.; Miyamoto, Y. Alveolar ridge preservation in beagle dogs using carbonate apatite bone substitute. *Ceram. Int.* **2022**, *48*, 1796–1804. [CrossRef]
14. Ogino, Y.; Ayukawa, Y.; Tachikawa, N.; Shimogishi, M.; Miyamoto, Y.; Kudoh, K.; Fukuda, N.; Ishikawa, K.; Koyano, K. Staged sinus floor elevation using novel low-crystalline carbonate apatite granules: Prospective results after 3-year functional loading. *Materials* **2021**, *14*, 5760. [CrossRef]
15. Kitamura, M.; Yamashita, M.; Miki, K.; Ikegami, K.; Takedachi, M.; Kashiwagi, Y.; Nozaki, T.; Yamanaka, K.; Masuda, H.; Ishihara, Y.; et al. An exploratory clinical trial to evaluate the safety and efficacy of combination therapy of REGROTH®and Cytrans®granules for severe periodontitis with intrabony defects. *Regen. Ther.* **2022**, *21*, 104–113. [CrossRef]
16. Tanimoto, K.; Sumi, K.; Yoshioka, M.; Oki, N.; Tanne, Y.; Awada, T.; Kato, Y.; Sugiyama, M.; Tanne, K. Experimental tooth movement into new bone area regenerated by use of bone marrow-derived mesenchymal stem cells. *Cleft Palate Craniofac. J.* **2015**, *52*, 386–394. [CrossRef]
17. Ten Heggeler, J.M.A.G.; Slot, D.E.; Van Der Weijden, G.A. Effect of socket preservation therapies following tooth extraction in non-molar regions in humans: A systematic review. *Clin. Oral Implant. Res.* **2011**, *22*, 779–788. [CrossRef]
18. Fukuba, S.; Okada, M.; Nohara, K.; Iwata, T. Alloplastic bone substitutes for periodontal and bone regeneration in dentistry: Current status and prospects. *Materials* **2021**, *14*, 1096. [CrossRef]
19. Fujioka-Kobayashi, M.; Katagiri, H.; Kono, M.; Schaller, B.; Iizuka, T.; Safi, A.F. The impact of the size of bone substitute granules on macrophage and osteoblast behaviors in vitro. *Clin. Oral Investig.* **2021**, *25*, 4949–4958. [CrossRef]
20. Maeno, S.; Niki, Y.; Matsumoto, H.; Morioka, H.; Yatabe, T.; Funayama, A.; Toyama, Y.; Taguchi, T.; Tanaka, J. The effect of calcium ion concentration on osteoblast viability, proliferation and differentiation in monolayer and 3D culture. *Biomaterials* **2005**, *26*, 4847–4855. [CrossRef]

21. Sato, N.; Handa, K.; Venkataiah, V.S.; Hasegawa, T.; Njuguna, M.M.; Yahata, Y.; Saito, M. Comparison of the vertical bone defect healing abilities of carbonate apatite, β-tricalcium phosphate, hydroxyapatite and bovine-derived heterogeneous bone. *Dent. Mater. J.* **2020**, *39*, 309–318. [CrossRef]
22. Canterelli, M.; Godoi, A.; Sinhoerti, M.; Neres, J.; Santos, E.; Sobrinho, L.; Costa, A. Effect of horizontal slot of maxillary canines' brackets with varying wire angulations—An in vitro study. *Braz. Dent. J.* **2022**, *33*, 55–63. [CrossRef]
23. Feller, L.; Khammissa, R.; Thomadakis, G.; Fourie, J.; Lemmer, J. Apical External Root Resorption and Repair in Orthodontic Tooth Movement: Biological Events. *Biomed. Res. Int.* **2016**, *2016*, 4864195. [CrossRef]
24. Ma, Z.; Wang, Z.; Zheng, J.; Chen, X.; Xu, W.; Zou, D.; Zhang, S.; Yang, C. Timing of force application on buccal tooth movement into bone-grafted alveolar defects: A pilot study in dogs. *Am. J. Orthod. Dentofacial. Orthop.* **2021**, *159*, e123–e134. [CrossRef]
25. Alalola, B.; Asiri, A.; Binmoghaiseeb, I.; Baharoon, W.; Alrassi, Y.; Alanizy, B.; Alsayari, H. Impact of Bone-Grafting Materials on the Rate of Orthodontic Tooth Movement: A Systematic Review. *Cureus* **2023**, *15*, e44535. [CrossRef]
26. Ru, N.; Liu, S.; Bai, Y.; Li, Y.; Liu, Y.; Wei, X. BoneCeramic graft regenerates alveolar defects but slows orthodontic tooth movement with less root resorption. *Am. J. Orthod. Dentofacial. Orthop.* **2016**, *149*, 523–532. [CrossRef]
27. Abe, T.; Sumi, K.; Kunimatsu, R.; Oki, N.; Tsuka, Y.; Awada, T.; Nakajima, K.; Sugiyama, M.; Tanimoto, K. Bone regeneration in a canine model of artificial jaw cleft using bone marrow–derived mesenchymal stem cells and carbonate hydroxyapatite Carrier. *Cleft Palate Craniofac. J.* **2020**, *57*, 208–217. [CrossRef]
28. Suzuki, S.; Venkataiah, V.S.; Yahata, Y.; Kitagawa, A.; Inagaki, M.; Njuguna, M.M.; Nozawa, R.; Kakiuchi, Y.; Nakano, M.; Handa, K.; et al. Correction of large jawbone defect in the mouse using immature osteoblast–like cells and a 3D polylactic acid scaffold. *PNAS Nexus.* **2022**, *1*, pgac151. [CrossRef]
29. Nakajima, K.; Kunimatsu, R.; Ando, K.; Ando, T.; Hayashi, Y.; Kihara, T.; Hiraki, T.; Tsuka, Y.; Abe, T.; Kaku, M.; et al. Comparison of the bone regeneration ability between stem cells from human exfoliated deciduous teeth, human dental pulp stem cells and human bone marrow mesenchymal stem cells. *Biochem. Biophys. Res. Commun.* **2018**, *497*, 876–882. [CrossRef]
30. Putranti, N.A.R.; Kunimatsu, R.; Rikitake, K.; Hiraki, T.; Nakajima, K.; Abe, T.; Tsuka, Y.; Sakata, S.; Nakatani, A.; Nikawa, H.; et al. Combination of carbonate hydroxyapatite and stem cells from human deciduous teeth promotes bone regeneration by enhancing BMP-2, VEGF and CD31 expression in immunodeficient mice. *Cells* **2022**, *11*, 1914. [CrossRef]

Disclaimer/Publisher's Note: The statements, opinions and data contained in all publications are solely those of the individual author(s) and contributor(s) and not of MDPI and/or the editor(s). MDPI and/or the editor(s) disclaim responsibility for any injury to people or property resulting from any ideas, methods, instructions or products referred to in the content.

Article

FEA Comparison of the Mechanical Behavior of Three Dental Crown Materials: Enamel, Ceramic, and Zirconia

Mario Ceddia, Luciano Lamberti * and Bartolomeo Trentadue

Dipartimento di Meccanica, Matematica e Management, Politecnico di Bari, 70125 Bari, Italy; m.ceddia@phd.poliba.it (M.C.); bartolomeo.trentadue@poliba.it (B.T.)
* Correspondence: luciano.lamberti@poliba.it

Abstract: The restoration of endodontically treated teeth is one of the main challenges of restorative dentistry. The structure of the tooth is a complex assembly in which the materials that make it up, enamel and dentin, have very different mechanical behaviors. Therefore, finding alternative replacement materials for dental crowns in the area of restorative care isa highly significant challenge, since materials such as ceramic and zirconia have very different stress load resistance values. The aim of this study is to assess which material, either ceramic or zirconia, optimizes the behavior of a restored tooth under various typical clinical conditions and the masticatory load. A finite element analysis (FEA) framework is developed for this purpose. The 3D model of the restored tooth is input into the FEA software (Ansys Workbench R23)and meshed into tetrahedral elements. The presence of masticatory forces is considered: in particular, vertical, 45° inclined, and horizontal resultant forces of 280 N are applied on five contact points of the occlusal surface. The numerical results show that the maximum stress developed in the restored tooth including a ceramic crown and subject to axial load is about 39.381 MPa, which is rather close to the 62.32 MPa stress computed for the natural tooth; stresses of about 18 MPa are localized at the roots of both crown materials. In the case of the zirconia crown, the stresses are much higher than those in the ceramic crown, except for the 45° load direction, while, for the horizontal loads, the stress peak in the zirconia crown is almost three times as large as its counterpart in the ceramic crown (i.e., 163.24 MPa vs. 56.114 MPa, respectively). Therefore, the zirconia crown exhibits higher stresses than enamel and ceramic that could increase in the case of parafunctions, such as bruxism. The clinician's choice between the two materials should be evaluated based on the patient's medical condition.

Keywords: dental stress analysis; finite element analysis; crown; dentin; crown materials; prosthetic dentistry

Citation: Ceddia, M.; Lamberti, L.; Trentadue, B. FEA Comparison of the Mechanical Behavior of Three Dental Crown Materials: Enamel, Ceramic, and Zirconia. *Materials* **2024**, *17*, 673. https://doi.org/10.3390/ma17030673

Academic Editors: Tobias Tauböck and Matej Par

Received: 29 December 2023
Revised: 24 January 2024
Accepted: 27 January 2024
Published: 30 January 2024

Copyright: © 2024 by the authors. Licensee MDPI, Basel, Switzerland. This article is an open access article distributed under the terms and conditions of the Creative Commons Attribution (CC BY) license (https://creativecommons.org/licenses/by/4.0/).

1. Introduction

Employing artificial crowns is a typical method in prosthetic dentistry for recreating the natural dental structure to solve problems, such as cavities and other structural injuries. Materials such as ceramics and metals have been very commonly used for prosthetic restoration, supported by natural teeth or implants [1,2]. Enamel makes up the natural tooth crown of the tooth, grinds food, and protects dentin, which acts as a force absorber during chewing. Dentin is a hard bone-like material that has an inner structure comprising a large number of tubules with variable diameter and spacing: this results in its anisotropic behavior. A study assessed the effect of various acids in cleaning the tooth surface, revealing that the use of polyacrylic acid is advantageous compared to other acids [3]. Some researchers [4] studied the influence of dentin tubules on mechanical characteristics. In particular, Kinney et al. [5] adopted a micromechanics-based approach to study the physical properties of dentin. Another approach followed in the literature was to take the transverse tubule of the dentin as a reference system and record the variation in mechanical properties along the tubule [6–8]. In particular, dentin was assessed to behave as a transversely isotropic material along the direction of the tubules (×1) (see Figure 1, taken from Ref. [8]).

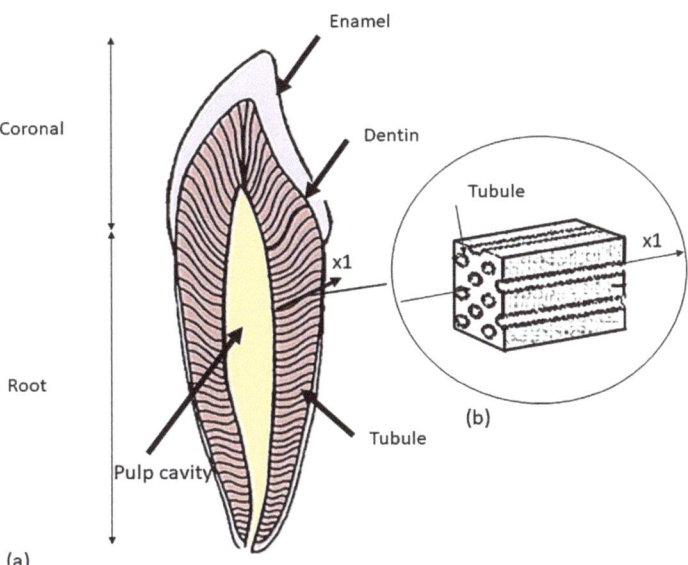

Figure 1. Schematic of a premolar tooth and the corresponding dentin microstructure: (**a**) tooth longitudinal section; (**b**) representation of the volume element extracted in the ×1 direction of the tubule.

In the late1990s, the achieved developments in the field of experimental techniques allowed the knowledge of dentin micromechanical behavior to be significantly improved [9]. For example, Wang et al. [10] used the intrinsic moiré structure to map strain distributions in the plane of the applied compressive load. It was found that dentin should be regarded as a non-homogeneous anisotropic material rather than homogeneous and isotropic. Kinney et al. [11] investigated the mechanical properties of dentin; in particular, the measured values of the Young's modulus were 30 GPa in the direction of the tubule and 15 GPa transversely to the tubule, with a Poisson's ratio ranging from 0.3 to 0.4. This suggests that dentin is stiffer along the direction of the tubule.

Enamel mechanically works to crush food during chewing and protects dentin thanks to its wear resistance. The elasticity of the underlying dentin prevents enamel from fracturing easily. However, because of its ectodermal origin, enamel does not possess vessels and cells; therefore, it cannot repair or grow once it has been secreted and matured. Hence, crack propagation due to oblique loads may cause chipping and damage in this layer. Therefore, when a replacement material for enamel is to be selected, the focus should be on its hardness and ability to absorb shocks. In recent years, many restorative materials, such as plastic (acrylic), metal, and porcelain, have been developed for dentistry applications, even though many patients prefer ceramic crowns because of their excellent biocompatibility, esthetics, and chemical durability [12].Their use has diffused since the 1990s, in spite of the fact that porcelain is a brittle material characterized by a very high risk of breakage. In order to solve this problem, porcelain was fused with metal, which prevented the formation of stress cracks [13]. Zirconia was later introduced to replace ceramics, due to its remarkable mechanical strength [14–16]. The excellent properties of zirconia derive from the phase variations occurring during heating and compaction processes. In fact, zirconia has three crystalline phases: a monoclinic phase at room temperature, which, at 1000 °C, turns into a tetragonal phase and, then, becomes stable above 2000 °C thanks to the addition of yttrium (YSZ) or magnesium (MSZ). This fundamental step was designed to preserve the crystalline structure when zirconia cools after being sintered at high temperatures. The tetragonal phase offers a greater mechanical strength than the monoclinic phase. In dentistry, yttrium-Stabilized zirconia (YSZ) is the most common formulation for the fabrication of dental restorations, such as crowns or bridges, due to its high strength (800–1200 MPa) and its ability to maintain its shape and size over time. In addition, during the cooling of zirconia, there is a 3–4% volume expansion that retards crack propagation (the fracture toughness increases to 6–8 MPa [14]).

The evaluation of the strength characteristics is a fundamental step in assessing the mechanical behavior of dental restorations. In particular, numerical approaches developed for dentin usually rely on homogenization. A representative volume element is extracted from the dentin tissue and its mechanical characteristics are studied using a micromechanics-based approach. Alternatively, analytical models can be used to solve the micromechanical problem. One of the most efficient models was developed by Mori-Tanaka et al. [17], who supplied an empirical model accounting for tubular variation, geometry, and spatial variation in the tubules. This model allowed a clearer understanding of the non-uniform behavior of dentin [17,18].

Modern dentistry carefully considers the application of occlusal forces, stress distribution, and strains because these factors significantly influence the success of the restoration. In this regard, many experimental approaches as well as numerical techniques have been developed over the years to study the distribution of stresses in restorative elements under masticatory loads. In particular, the finite element method (FEM) has been shown to be a valid complement/alternative to the experimental assessment of the biomechanical behavior of restored teeth [19]. FEM solves an approximate problem defined by discretizing a geometric model describing the domain of the problem at hand into a 3D mesh comprising a finite number of elements of finite size and simple geometry. Elements are connected by characteristic points called nodes at which the structural response is computed. Each element of the FE model is subject to a set of applied loading conditions and kinematic constraints with the aim of deciding the global behavior of the discretized body.

Earlier studies published in the technical literature employed FEM to investigate the 2D biomechanical behavior of enamel, modeling this material as either isotropic or anisotropic [20] or purely isotropic [21]. Other studies later evaluated the biomechanical behavior of ceramic and zirconia restorative materials used for replacing natural enamel [22,23]. The present study aims to evaluate, using finite element analyses (FEA), the 3D biomechanical behavior of premolar teeth including restored crowns under axial and inclined occlusal loads. For that purpose, the stress distributions of restored teeth including ceramics or zirconia, are compared to those determined for natural teeth including isotropic enamel. The novelty of this study consists in the comparison between the two restorative materials and enamel, as all three materials are studied simultaneously. Additionally, the anisotropic behavior of enamel enhances accuracy. The bone region into which the premolar tooth is inserted is also modeled, and changes in stress distributions in the bone tissues are determined for the two restorative materials with respect to the case of natural teeth. The null hypothesis stating that varying crown material does not affect the results in terms of stress/deformation is not confirmed.

The rest of the article is structured as follows. Section 2 describes in detail the FE modeling process of the premolar tooth. Section 3 presents the FE solution options and the numerical results obtained for the different combinations of occlusal loads and materials. Section 4 discusses the results of Section 3 in the context of the technical literature. Section 5 summarizes the main findings and highlights the limitations and directions of future investigations.

2. Materials and Methods

2.1. CAD Model

The computer-aided design (CAD) model was obtained from computed tomography (CT) scans of real premolar teeth, also considering the modeling conducted by Yoon et al. [24]. Recomposition and layering were processed using the Autodesk Inventor 2023®CAD (2023, San Francisco, CA 94105, USA) environment. The scan file coded in the .STL format was later converted into the .STP format. The thickness of the crown was selected from the recently published literature [24], where the thickness of the enamel layer was indicated to range from 0.3 mm to 2.5 mm (see Figure 2).

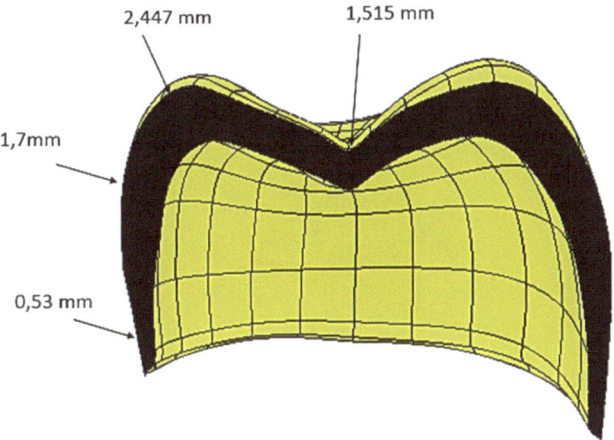

Figure 2. Cross-sectional view of the crown, showing the spatial variations in the crown thickness.

Subsequently, a 3D model of the premolar tooth suitable for accommodating the crown was prepared, which was then placed on the dentin [25,26], as it is shown in Figure 3.

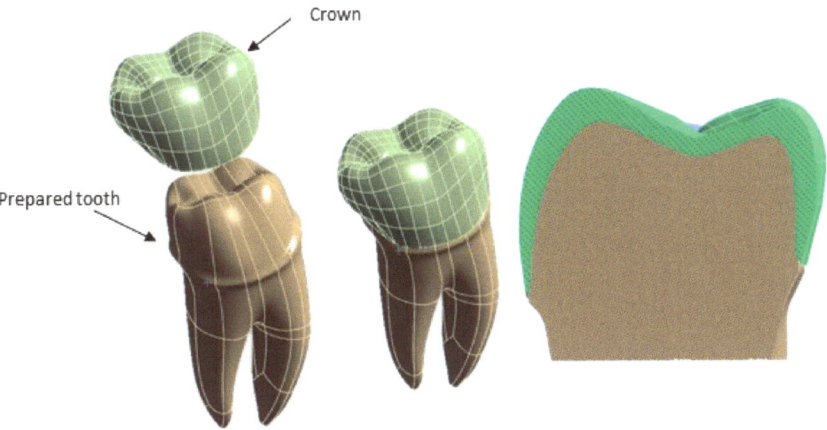

Figure 3. Components of the 3D model of the premolar tooth to be analyzed using FEA.

In order to make the FE model of the premolar tooth more reliable, a cylinder was created that simulated bone (cortical and trabecular) with a diameter of 15 mm and a height of 20 mm with a cortical bone thickness of 2 mm [27,28] (see Figure 4).

2.2. Finite Element Modeling

The CAD file of the 3D model of the premolar tooth was exported in the STP format and then input into the ANSYS Workbench 2023® FEA software (R23, Canonsburg, PA, USA). This model was mainly discretized with tetrahedral elements (SOLID 187). In order to find the best trade-off between the accuracy of the results and the computational cost of FEA, a mesh convergence analysis was conducted by selecting the maximum principal stress as the target quantity. The element size was progressively reduced until the last three values of principal stress differed by less than 1%.

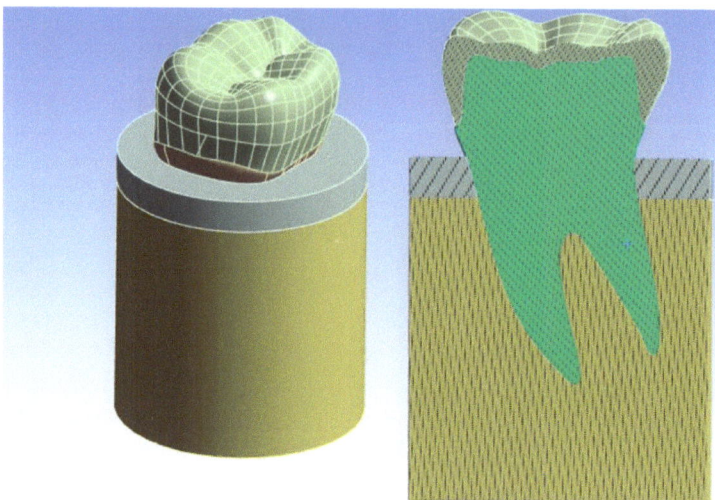

Figure 4. Overall model of the tooth inserted into the bone.

Figure 5 shows that the maximum principal stress tends to reach an asymptotic value over a plateau starting at the red point denoted by about 140,000 elements. An element size of 0.8 mm was, therefore, chosen in this study for the finite element analyses. The details of the final meshes of the different parts of the FE model of the premolar tooth are presented in Figure 6.

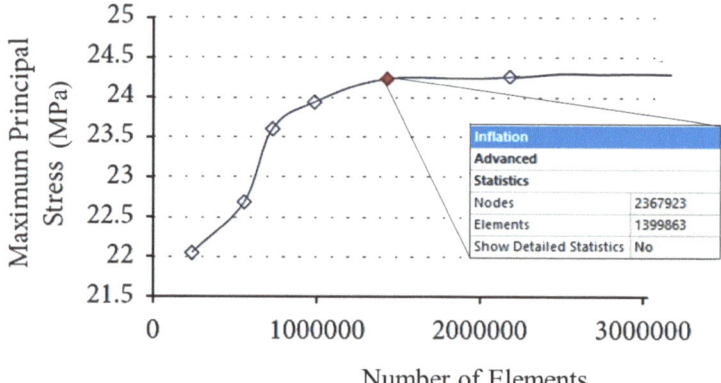

Figure 5. Results of the mesh sensitivity analysis.

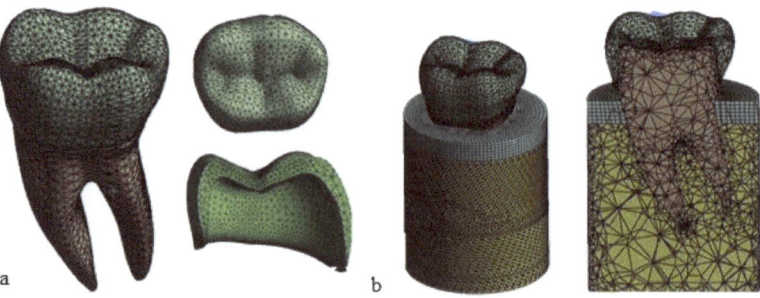

Figure 6. Mesh view display: (**a**) tooth and crown; (**b**) complete FE model of the tooth with the trabecular bone in yellow and the cortical bone in gray (element size = 0.8 mm).

2.3. Material Properties

The shaped tooth can be divided into three regions: crown, cement layer, and dentin. Three materials were selected in this study for the crown: enamel (natural tooth), ceramic, and zirconia (restored tooth). As for the bone, it can be divided into cortical bone and cancellous bone. Their difference is mainly due to their mechanical characteristics. In particular, the cortical bone is denser and has a better mechanical behavior than the trabecular bone. The presence of internal trabeculae leads to a reduced density and, hence, lower mechanical characteristics in the case of the trabecular bone [29,30]. To obtain information on the bone mechanical properties (i.e., Young's modulus and Poisson's ratio) and consequently on the deformation behavior, one can rely on mathematical relationships correlating the mass density of CT-scanned elements characterized by values expressed in Hounsfield units. These values can be related to the density of the element using Equation (1) [31].

$$\rho = 0.007764 HU - 0.05614 \tag{1}$$

Wirtz et al. [32] proposed a mathematical relationship between the modulus of elasticity and density of the cortical and trabecular bones. That is:

$$Ecort = 2.065 \times \rho^{3.09} \tag{2}$$

$$Etrab = 1.904 \times \rho^{1.64} \tag{3}$$

The method described in [32] assumes that the bone material has an isotropic behavior with the same thermo-mechanical characteristics in all directions. However, other studies [33–35] proved that the bone response to external loads is best described by an anisotropic behavior (i.e., the thermo-mechanical characteristics are different in all directions) due to the non-homogeneity of the material because of the presence of trabeculae and the different responses to tensile and compressive loads. In view of this, the bone tissues were also modeled as anisotropic materials in this study using the mechanical properties listed in Table 1 [36].

Table 1. Anisotropic mechanical properties of the bone tissues input into the FEA of the restored teeth.

Material	E_z (GPa)	E_y (GPa)	E_x (GPa)	ν_{xy}	ν_{yx}	ν_{xz}	G_{xy} (GPa)	G_{yz} (GPa)	G_{xz} (GPa)
Cortical bone	17.9	12.5	26.6	0.28	0.18	0.31	7.1	4.5	5.3
Cancellous bone	0.21	1.148	1.148	0.055	0.322	0.055	0.068	0.434	0.068

As mentioned in the Introduction, dentin has an anisotropic behavior due to the microscopic nature of the tissue, which is formed by a set of tubules that confer anisotropy along the longitudinal directions of the tubules themselves. Table 2 lists the anisotropic properties of dentin input into the finite element model [37].

Table 2. Anisotropic mechanical properties of dentin input into the FEA of the restored teeth.

Material	E_z (GPa)	E_y (GPa)	E_x (GPa)	ν_{xy}	ν_{yx}	ν_{xz}	G_{xy} (GPa)	G_{yz} (GPa)	G_{xz} (GPa)
Dentin	17.07	5.61	5.61	0.30	0.33	0.30	1.7	6	1.7

Munari et al. [38] compared the isotropic and anisotropic mechanical behaviors of natural enamel, finding marginal differences between the results obtained for these two hypotheses. Table 3 lists the modulus of elasticity E and the Poisson's ratio values (the same values in all directions) used in this study [39].

Ceramic and zirconia restorative materials were also modeled as isotropic materials in this study on the basis of the data reported in the technical literature [40,41]. The corresponding mechanical properties used for the FEA are listed in Table 4.

Table 3. Isotropic mechanical properties of natural enamel input into the FEA of the restored teeth.

Material	E (GPa)	ν
Enamel	72.7	0.30

Table 4. Crown materials properties.

Material	E (GPa)	ν
Zirconia	205	0.22
Porcelain	68.9	0.28

Table 5 shows the strength limits of the materials studied in this paper. The listed values were extracted from [42–45].

Table 5. Tensile strength and compressive strength of the analyzed materials.

Material	Tensile Strength (MPa)	Compressive Strength (MPa)
Enamel	11.5	384.0
Dentin	105.5	297.0
Zirconia	745.0	2000.0
Ceramic		330
Cortical bone	135	205
Trabecular bone		10.44

2.4. Loads and Constraints

Three loading directions were considered to simulate mastication forces: a vertical (axial) load, an angled (45°) load, and a horizontal load. The loads were applied at five points on the occlusal surface of the dentin. These points simulated the possible points of contact during chewing. The intensity of the resultant load applied to the tooth was 280 N [40]. Figure 7 summarizes the applied forces and where they act on the tooth structure.

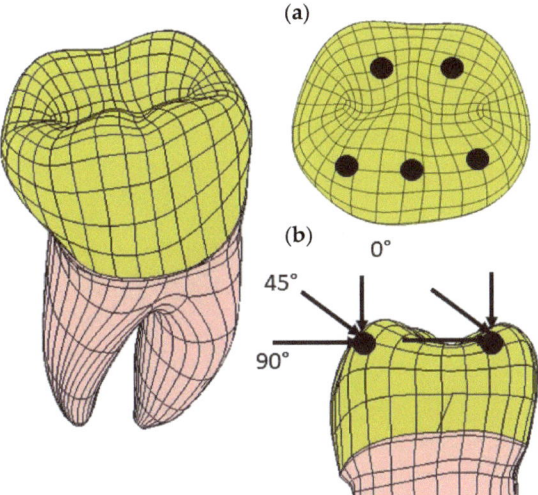

Figure 7. Load application points and the corresponding loading directions of the mastication forces simulating the maximum bite force: (**a**) three points are located on the inclined external sides of each buccal cusp and two points are located on the inclined internal sides of each lingual cusp; (**b**) there are three different load directions: vertical (axial), oblique (45°), and horizontal.

2.5. Kinematic Constraint Conditions

The lateral surfaces and the lower surface of the cylinder simulating the presence of the bone regions were fixed in all directions. To simulate a perfect osseointegration between tooth and bone, a fixed contact condition was selected (see Figure 8a). In addition, a fixed frictionless contact condition was also selected for the crown/dentin interface (see Figure 8b).

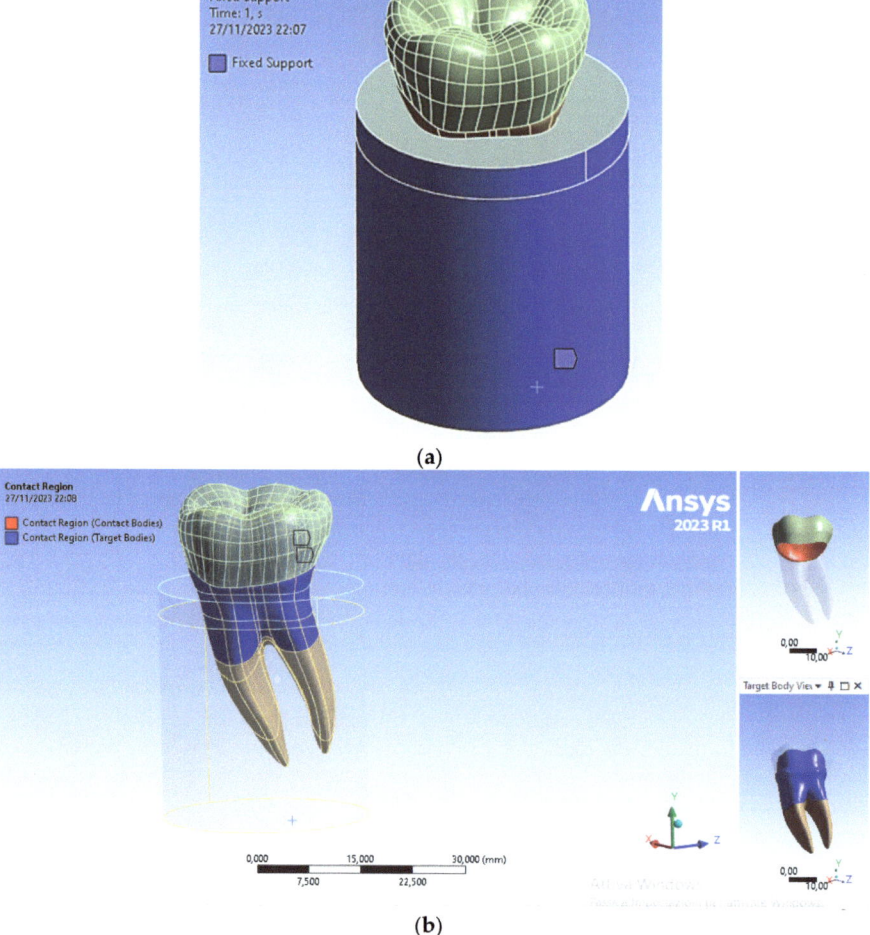

Figure 8. Boundary and contact conditions selected in the FEA of the restored premolar tooth: (**a**) fixed surfaces to simulate the rigid support provided by the bone region to the tooth; (**b**) type of contact between dentin and crown.

3. Results

The distributions of Von Mises equivalent stress and the corresponding stress peaks were the main output quantity obtained from FEA. The Von Mises stress supplies a single measure of the equivalent stress that, if exceeded, may yield the initiation of plastic deformation in the material. Stress distributions were plotted using ANSYS in the fashion of 3D maps with colors ranging from blue (low stress) to red (high stress).This allowed high stress regions to be promptly identified and, hence, to understand the mechanical behavior of the restored teeth under the occlusal loads shown in Figure 6 (i.e., 280 N resultant force acting in the apical direction, lingual, and 45° inclined direction applied to the occlusal surface of the crown).

Figure 9 shows the Von Mises stress distribution computed using FEA for the natural tooth subject to a 280 N occlusal load acting in the vertical direction. Figures 10 and 11, respectively, show the Von Mises stress distributions obtained using ANSYS for the 45° inclined load and the horizontal load acting on the premolar tooth.

It can be seen that applying the load to the tooth in the apical direction (Figure 9) leads to a maximum Von Mises stress of 62.32 MPa in the cervical part of the tooth. Stresses in the occlusal surface range from 3 to 15 MPa with localized peaks in the grooves. Furthermore, looking at Figure 9, it can be seen that the occlusal load is equally distributed between two roots, which are stressed in the same way by a stress of about 3 MPa, with a concentration of stress in the area corresponding to the cervical part of the crown.

Figure 10 shows how the 45° inclined load generates load components in the direction perpendicular to the apical direction as well. This causes Von Mises stress to increase up to about 138 MPa in the cervical area. However, the stress distribution becomes asymmetric at the tooth roots. The right root that is in the direction of the inclined occlusal force is more stressed, reaching a maximum stress of about 4 MPa vs. only about 3.2 MPa localized in the left root. Figure 11 demonstrates that the application of a load parallel to the occlusal surface generates a maximum equivalent stress of about 91.465 MPa. However, unlike the other two load configurations previously analyzed, in this case, the occlusal surface of the crown is more stressed: a 58.32 MPa stress peak vs. only 12 MPa (vertical load) and 15 MPa (45° inclined load).

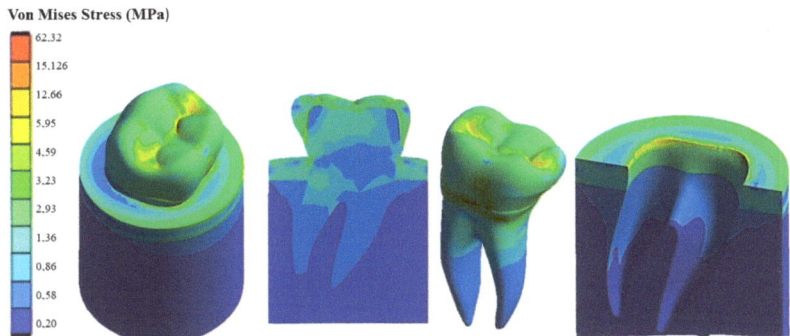

Figure 9. Von Mises stress distribution computed using FEA for a natural tooth (enamel crown) subject to an axial occlusal force.

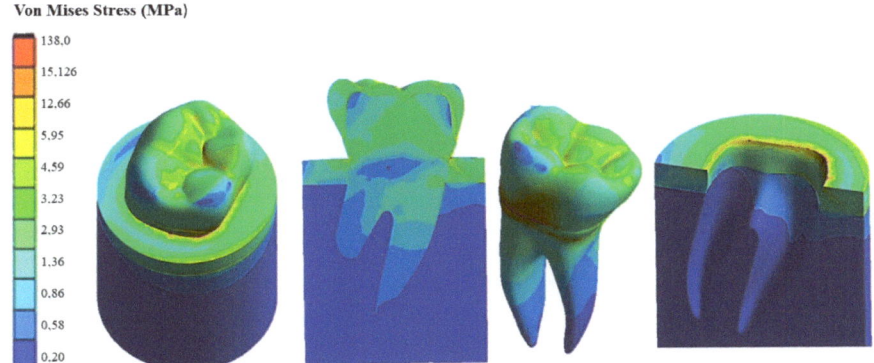

Figure 10. Von Mises stress distribution computed using FEA for a natural tooth (enamel crown) subject to occlusal forces inclined by 45°.

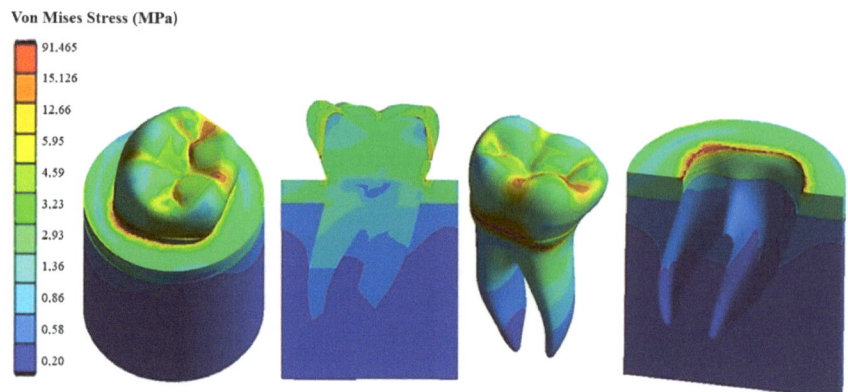

Figure 11. Von Mises stress distribution computed using FEA for a natural tooth (enamel crown) subject to a horizontal occlusal force.

As expected, the cortical bone region had a higher stress than the spongy bone region in all load configurations because of its higher stiffness. This generated a phenomenon of stress shielding in the trabecular bone. The highest stress at the tooth–bone interface was observed for the inclined loading configuration: about 76 MPa vs. only 46 MPa computed for the axial load configuration and only 53 MPa computed for the horizontal load.

Figure 12 presents the Von Mises stress distribution computed using ANSYS for the restored tooth including ceramic as a replacement material for the crown. In the case of axial load (see Figure 12a), the maximum stress developed in the model is lower than that determined for the natural tooth (i.e., only 39.381 MPa vs. 62.32 MPa) and tends to be homogeneously distributed over the entire occlusal surface. At the roots, the stress is similar to its counterpart determined for the natural tooth (about 18 MPa). By inclining the force to 45° (see Figure 12b), the equivalent stress increased over the whole tooth structure in an analogous way to the case of the natural tooth with an enamel crown, yet it remained considerably lower: only about 55.691 MPa vs. about 138 MPa, respectively. Finally, the horizontal load (see Figure 12c) resulted in only a slight increase in the stress with respect to the case of the 45° inclined load, only 56.114 MPa vs. 55.691 MPa, respectively, while in the case of the natural tooth, the maximum stress varied to a large extent, dropping from about 138 MPa to about 91.465 MPa.

Figure 13 presents the FEA results obtained for the restored premolar with a zirconia crown. In the case of axial and horizontal loadings, the computed stresses in the tooth structure were significantly higher than those computed for both the restored premolar tooth with a ceramic crown and the natural tooth with an enamel crown. The stress values computed for the occlusal load inclined at 45° were similar for both restored teeth. Figure 14 summarizes the main results obtained in terms of the maximum stress whose values are reported in the graph for the different combinations of tooth structure (i.e., natural, restored with a ceramic crown, and restored with a zirconia crown) and loading conditions (axial, 45° inclined, and horizontal forces). All stress values computed using ANSYS were lower than the strength limits indicated in Table 5. Figure 14 shows that using the ceramic crown restoration allows us to obtain a similar mechanical behavior under all loading conditions experienced by the tooth structure (i.e., axial, inclined, and horizontal forces). Conversely, the zirconia-crown-restored tooth shows significant stress peaks for perfectly axial and horizontal loads. In terms of the stress level and the consequent risk of fracture, the ceramic crown restoration may require more attention to avoid physical damage, while the zirconia crown restoration may be more tolerant to mechanical stresses. This is confirmed in Figure 12: in the case of the ceramic restoration, the maximum stresses are localized at certain critical points and such a non-uniform distribution may increase the probability of initiating fracture phenomena.

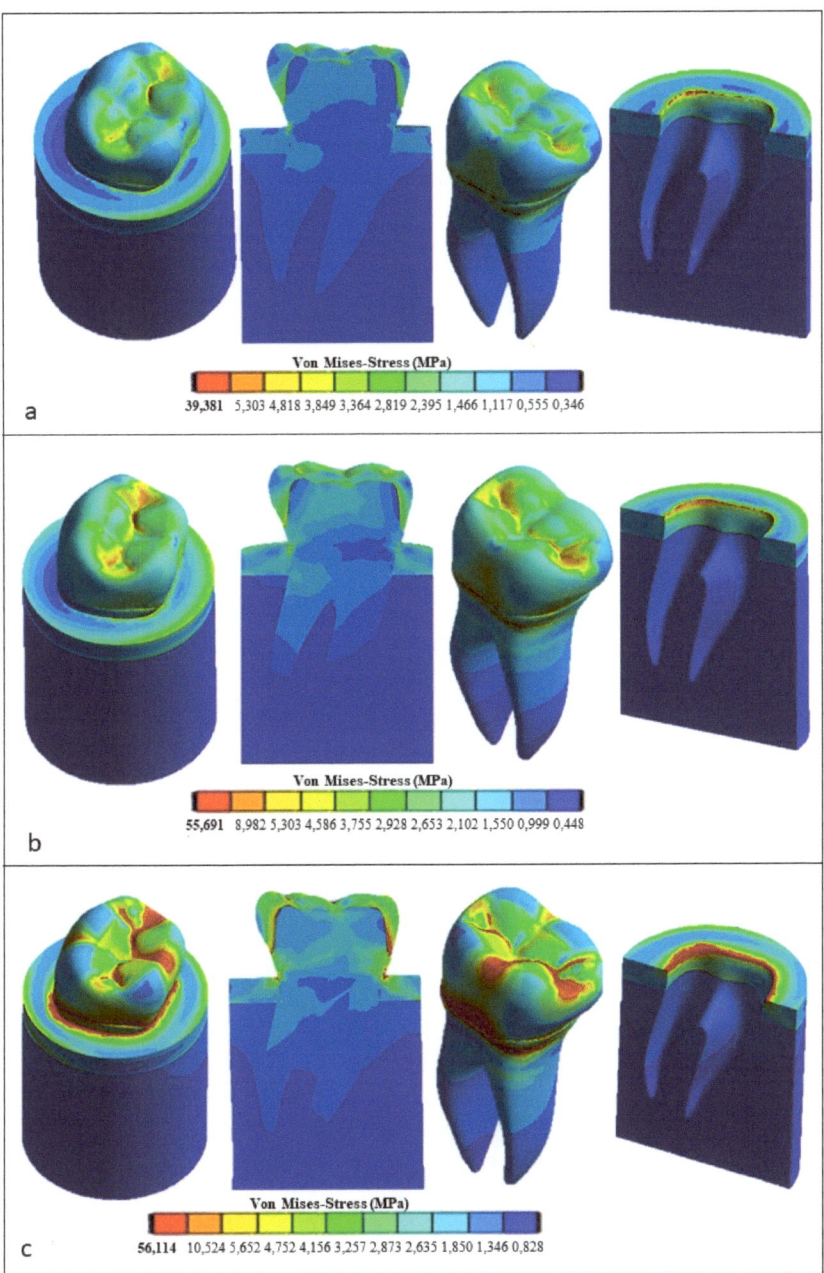

Figure 12. Von Mises stress distributions computed using FEA for the restored tooth with a ceramic crown: (**a**) vertical load, (**b**) 45° inclined load, and (**c**) horizontal load.

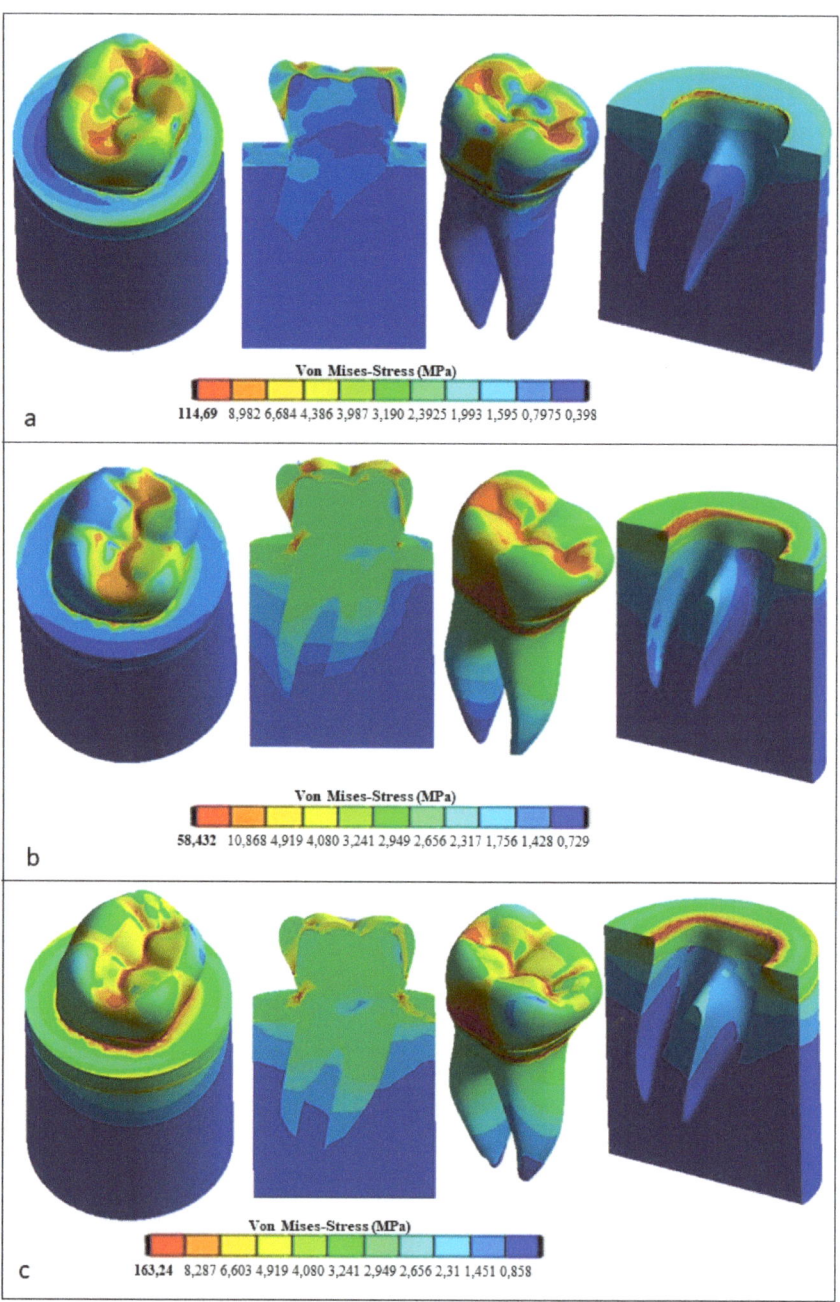

Figure 13. Von Mises stress distributions computed using FEA for the restored tooth with a zirconia crown: (**a**) vertical load; (**b**) 45° inclined load; and (**c**) horizontal load.

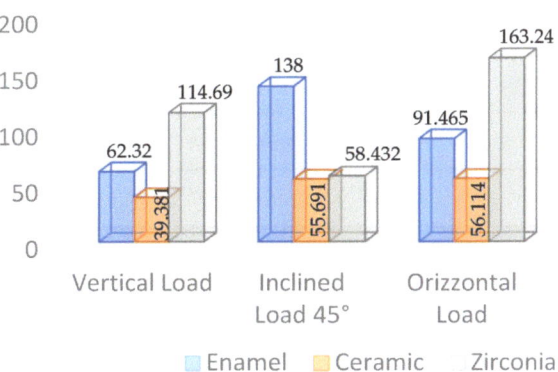

Figure 14. Maximum Von Mises stress values computed using FEA for the different dental structures under various loading conditions.

4. Discussion

The present study investigated the effect of the dental crown material (enamel for the natural tooth and ceramic and zirconia crowns for the restored teeth) on the 3D stress distribution of a mandibular premolar tooth. Three-dimensional finite element models, including the tooth structure and a cylindrical block consisting of the cortical bone and trabecular bone to support the tooth, were developed for this study. The zirconia restoration was selected because its mechanical properties provide excellent mechanical strength in dental applications. These characteristics can be greatly improved by firing at a high temperature, which triggers transformation hardening that is opposed to the propagation of cracks [46]. Therefore, the zirconia restoration has significantly higher mechanical properties than other restorative materials, like ceramics [47,48]. In [49], it was shown that zirconia is stable at 1170 °C, but it has a cubic structure at 2370 °C [50]. Most zirconia-based prosthetic structures are made of yttrium-stabilized zirconium polycrystals (3Y-TZP) [48]. The most important advantage of this stabilizer is its high fracture toughness and flexural strength [51]. Despite the popularity of zirconia, various complications have been reported in the technical literature for this restorative material [52–54]. For example, there can be residual thermal stresses due to mismatches between the coefficient of thermal expansion and differences in the modulus of elasticity between zirconia and the coating material, which may facilitate a chipping fracture [55,56].

Other clinical trial data [57–59] focused on the fact that the type of fixation agent between the crown and abutment can affect the retention of zirconium-based crowns and the durability of the implant. Furthermore, the thickness of the cement layer significantly influences crown retention. If the cement layer is thin, the gap between the prosthetic crown and abutment may become very small, thus increasing the forces needed to extract the prosthetic crown from the abutment and compromising the durability of the prosthetic application [58]. Ceramics are brittle materials and, hence, very susceptible to a risk of fractures [53]. To reduce this risk, the ceramic is melted with metal alloys that provide a certain toughness to the structure. Aceramic crown, compared to the zirconia one, provides a very natural and translucent appearance, similar to that of the natural tooth. Even though it is esthetically better, ceramic is less resistant to fracture than zirconia [60]. In addition, even when fitting with a natural tooth, the ceramic crown does not require a removal of the tooth structure that is as considerable as that with zirconia, as the ceramic is to be thinner.

One approach to investigating the integrity of dental structures is to use destructive tests that apply load cycles on the tooth element through a ball or bar [61]. This in vitro method has limitations because the testing machine setup may not directly simulate the actual oral conditions resulting from variations in masticatory load as well as the directionality of the load. Sorrentino et al. [62] evaluated in vitro the mechanical strength of a monolithic zirconia crown obtaining about 1655 N vs. only about 1400 N for ceramics. This confirmed the higher mechanical strength of zirconia in the case of masticatory loads. FEA was used to mimic the intraoral conditions to evaluate the fracture strength of various materials used in dental restorations [63–66]. Alsadon et al. [51] obtained similar results for zirconium-coated crowns and zirconium–porcelain-composite-coated crowns: the peak stresses were, respectively, equal to 63.6 MPa and 50.9 MPa. Fathy et al. [67] studied the finite element stress distribution in fully milled or layered zirconia crowns. They found that the single-material zirconia-restored crown was stiffer than the layered crown restoration. Other FEA studies [67–69] also investigated stress distribution in bone based on the selected material for the crown: the porcelain coating led to a reduction in stresses in the bone due to the lower elastic modulus compared to zirconia.

The results obtained in present paper are consistent with those reviewed above. In fact, stiffness variations in the restorative material used for replacing enamel may result in a significant stress reduction in the tooth elements. Such an effect is more pronounced if the horizontal components of the occlusal loads predominate, thus stressing the occlusal surface. Angular and horizontal loads cause the stress distribution to become wider than in the case of the vertical load. These loads are generated more in the cervical region than in the apical region of the tooth.

Some studies [70] evaluating cement spaces demonstrated how the presence of a larger cement layer localizes peak stresses in marginal areas of the concrete, making it more susceptible to failure. In this study, the cement layer was omitted to study precisely the critical condition in which there is a higher contact stress between the crown and the prosthetic system. In addition, it was seen that, compared to natural enamel, the ceramic crown generated less stress in the bone than the zirconia crown. A prosthetic restoration with an osseointegrated implant in the bone is currently being developed also considering that stresses in the bone that are very low may cause bone resorption. Therefore, zirconia turns out to be the best replacement material of natural enamel in terms of optimizing osteointegration at the tooth–bone interface.

A limitation of the present study is in the modeling of the contact at the tooth–bone interface. In this paper, the modeling of periodontal ligament (PDL) was omitted because several studies [71–73], focusing on the influence that the periodontal ligament can have on the stress transfer from teeth to the bone, indicated such influence to become significant only if the study focuses exclusively on the tooth–bone contact. Conversely, if the research scope is to investigate the overall mechanical behavior of the tooth structure, the presence of the periodontal ligament can be omitted. Moreover, since the mechanical behavior of the periodontal ligament is not yet fully understood, it is very difficult to reliably model it in the context of FEA. Controversial studies [74,75] indicated a hyperplastic behavior is more suited than a viscoelastic behavior for the periodontal ligament. Other studies [76,77] stated that varying the PDL's constitutive behavior may change the position of the tension peaks, which can be translated from the cervical area in the case of the PDL's hyperelastic behavior to the distal area if a viscoelastic or elastoplastic behavior is assumed for the periodontal ligament. The relative limitations of this study mainly involve the simplification of the model, as it was not possible to model all components, such as the periodontal ligament, due to the limited availability of studies allowing the PDL's mechanical characterization. Additionally, the dependence on input data, such as loads and constraints, may have led to less accurate results. The mechanical properties of the materials used in the simulation significantly influence the outcomes. There are studies considering isotropic or anisotropic bone, and different behaviors may lead to the creation of anomalous stress fields at the bone–tooth interface. Therefore, FEA studies should be compared with in vitro tests to ensure result accuracy.

5. Conclusions

In this study, FEA confirmed itself as an extremely useful tool for evaluating stresses in a complex biomechanical structure comprising different materials, such as a restored mandibular premolar tooth. Using 3D modeling and numerical analyses, it was possible to understand how the selection of the crown restorative material affects the stress distribution with respect to the natural tooth. Despite having excellent esthetic characteristics, ceramic has a lower resistance to occlusal loads than zirconia. It was seen that varying the stiffness of the selected crown replacement materials (i.e., zirconia and ceramics) significantly affected the stress distribution in the restored tooth. Occlusal load direction also affected the intensity and distribution of the transmitted stress. In particular, the application of a horizontal load significantly increased the stress on the occlusal surface of the zirconia crown with respect to the ceramic crown restoration, which appeared to be rather insensitive to the direction of applied force. The effect of the cement layer between the crown and dentin was not considered because the focus was more on the analysis of the occlusal surface of the tooth. The stress at the tooth–bone interface was also influenced by the presence of the periodontal ligament, which, due to the general analysis of transmitted stress, was not considered in this study. Future investigations should include the correct implementation of these elements in order to more accurately assess the transmitted stresses through the tooth structure interfaces.

Author Contributions: Conceptualization, M.C. and B.T.; methodology, M.C.; software, M.C.; validation, B.T. and L.L.; formal analysis, M.C.; investigation, B.T. and L.L.; resources, B.T.; data curation, M.C.; writing—original draft preparation, M.C. and B.T.; writing—review and editing, M.C., B.T. and L.L.; visualization, B.T.; supervision, B.T.; project administration, L.L. and B.T. All authors have read and agreed to the published version of the manuscript.

Funding: This research received no external funding.

Institutional Review Board Statement: Not applicable.

Informed Consent Statement: Not applicable.

Data Availability Statement: All experimental data to support the findings of this study are available from the corresponding authors upon request.

Conflicts of Interest: The authors declare no conflicts of interest.

References

1. Alhasanyah, A.; Vaidyanathan, T.K.; Flinton, R.J. Effect of core thickness differences on post-fatigue indentation fracture resistance of veneered zirconia crowns. *J. Prosthodont.* **2013**, *22*, 383–390. [CrossRef]
2. Christensen, G.J. Porcelain-fused-to-metal versus zirconia based ceramic restorations. *J. Am. Dent. Assoc.* **2009**, *140*, 1036–1039. [CrossRef] [PubMed]
3. Kharouf, N.; Mancino, D.; Naji-Amrani, A.; Eid, A.; Haikel, Y.; Hemmerle, J. Effectiveness of etching by three acids on the morphological and chemical features of dentin tissue. *J. Contemp. Dent. Pract.* **2019**, *20*, 915–919.
4. Ogata, M.; Okuda, M.; Nakajima, M.; Pereira, P.N.; Sano, H.; Tagami, J. Influence of the direction of tubules on bond strength to dentin. *Oper. Dent.* **2001**, *26*, 27–35.
5. Kinney, J.H.; Balooch, M.; Marshall, S.J.; Marshall, W.J.R.; Weih, T.P. Hardness and Young's modulus of human peritubular and intertubular dentine. *Arch. Oral Biol.* **1996**, *41*, 9–13. [CrossRef] [PubMed]
6. Wentrup-Byrne, E.; Armstrong, C.A.; Armstrong, R.S.; Collins, M. Fourier transform Raman microscopic mapping of the molecular components in a human tooth. *J. Raman Spectrosc.* **1997**, *28*, 151–158. [CrossRef]
7. Huo, B.; Zheng, Q.S. Effect of dentin tubules to the mechanical properties of dentin. Part I: Stress–strain relations and strength criterion. *Acta Mech. Sin. (Engl. Ser.)* **1999**, *15*, 355–364.
8. Lertchirakarn, V.; Palamara, E.A.; Messer, H.H. Anisotropy of tensile strength of root dentin. *J. Dent. Res.* **2001**, *80*, 453–456. [CrossRef] [PubMed]
9. Huo, B.; Zheng, Q.S.; Zhang, Q.; Wang, J.D. Effect of dentin tubules to the mechanical properties of dentin. Part II: Experimental study. *Acta Mech. Sin. (Engl. Ser.)* **2000**, *16*, 75–82.

10. Wang, R.Z.; Weiner, S. Strain–structure relations in human teeth using moiré fringes. *J. Biomech.* **1998**, *31*, 135–141. [CrossRef]
11. Kinney, J.H.; Balooch, M.; Marshall, G.W.; Marshall, S.J. A micromechanics model of the elastic properties of human dentine. *Arch. Oral Biol.* **1999**, *44*, 813–822. [CrossRef]
12. Pol-Christian, W.P.; Kalk, W. A systematic review of ceramic inlays in posterior teeth: An update. *Int. J. Prosthodont.* **2011**, *24*, 566–575.
13. Asgar, K. Casting metals in dentistry: Past present future. *Adv. Dent. Res.* **1988**, *23*, 3–43. [CrossRef]
14. Jiang, L.; Zhao, Y.-Q.; Zhang, J.-C.; Liao, Y.-M.; Li, W. Zhonghua kou qiang yi xue za zhi = Zhonghua kouqiang yixue zazhi. *Chin. J. Stomatol.* **2010**, *45*, 376–380.
15. Kosmac, T.; Oblak, C.; Jevnikar, P.; Funduk, N.; Marion, L. The effect of surface grinding and sandblasting on flexural strength and reliability of Y-TZP zirconia ceramic. *Dent. Mater.* **1999**, *15*, 426–433. [CrossRef]
16. Blatz, M.B.; Sadan, A.; Arch, G.H., Jr.; Lang, B.R. In vitro evaluation of long-term bonding of ProceraAllCeram alumina restorations with a modified resin luting agent. *J. Prosthet. Dent.* **2003**, *89*, 381–387. [CrossRef]
17. Mori, T.; Tanaka, K. Average stress in matrix and average elastic energy of materials with misfitting inclusions. *Acta Metall.* **1973**, *21*, 571–574. [CrossRef]
18. Gommers, B.; Verpoest, I.; Van Houtte, P. The Mori-Tanaka method applied to textile composite materials. *Acta Mater.* **1998**, *46*, 2223–2235. [CrossRef]
19. Grzebieluch, W.; Będziński, R.; Czapliński, T.; Kaczmarek, U. The mechanical properties of human dentin for 3-D finite element modeling: Numerical and analytical evaluation. *Adv. Clin. Exp. Med.* **2017**, *26*, 645–653. [CrossRef] [PubMed]
20. Shinno, Y.; Ishimoto, T.; Saito, M.; Uemura, R.; Arino, M.; Marumo, K.; Nakano, T.; Hayashi, M. Comprehensive analyses of how tubule occlusion and advanced glycation end-products diminish strength of aged dentin. *Sci. Rep.* **2016**, *6*, 19849. [CrossRef]
21. Reddy, M.K.; Vandana, K.L. Three-dimensional finite element analysis of stress in the periodontium. *J. Int. Acad. Periodontol.* **2005**, *7*, 102–107.
22. Bramanti, E.; Cervino, G.; Lauritano, F.; Fiorillo, L.; D'amico, C.; Sambataro, S.; Denaro, D.; Famà, F.; Ierardo, G.; Polimeni, A.; et al. FEM and Von Mises analysis on prosthetic crowns structural elements: Evaluation of different applied materials. *Sci. World J.* **2017**, *2017*, 1029574. [CrossRef] [PubMed]
23. Kanat, B.; Cömlekoğlu, E.M.; Dündar-Çömlekoğlu, M.; Hakan Sen, B.; Özcan, M.; Güngör, M.A. Effect of various veneering techniques on mechanical strength of computer-controlled zirconia framework designs. *J. Prosthodont.* **2014**, *23*, 445–455. [CrossRef] [PubMed]
24. Yoon, Y.; Lee, M.-J.; Kang, I.; Oh, S. Evaluation of Biomechanical Stability of Teeth Tissue According to Crown Materials: A Three-Dimensional Finite Element Analysis. *Materials* **2023**, *16*, 4756. [CrossRef]
25. Shahmoradi, M.; Wan, B.; Zhang, Z.; Wilson, T.; Swain, M.; Li, Q. Monolithic crowns fracture analysis: The effect of material properties, cusp angle and crown thickness. *Dent. Mater.* **2020**, *36*, 1038–1051. [CrossRef]
26. Ha, S.R. Biomechanical three-dimensional finite element analysis of monolithic zirconia crown with different cement type. *J. Adv. Prosthodont.* **2015**, *7*, 475–483. [CrossRef]
27. Di Pietro, N.; Ceddia, M.; Romasco, T.; De Bortoli Junior, N.; Mello, B.F.; Tumedei, M.; Specchiulli, A.; Piattelli, A.; Trentadue, B. Finite element analysis (FEA) of the stress and strain distribution in cone-morse implant–abutment connection implants placed equicrestally and subcrestally. *Appl. Sci.* **2023**, *13*, 8147. [CrossRef]
28. Ceddia, M.; Comuzzi, L.; Di Pietro, N.; Romasco, T.; Specchiulli, A.; Piattelli, A.; Trentadue, B. Finite Element Analysis (FEA) for the evaluation of retention in a Conometric Connection for Implant and Prosthesis. *Osteology* **2023**, *3*, 140–156. [CrossRef]
29. Li, P.; Shen, L.; Li, J.; Liang, R.; Tian, W.; Tang, W. Optimal design of an individual endoprosthesis for the reconstruction of extensive mandibular defects with finite element analysis. *J. Cranio-Maxillofac. Surg.* **2014**, *42*, 73–78. [CrossRef] [PubMed]
30. Van Eijden, T.M. Biomechanics of the mandible. *Crit. Rev. Oral Biol. Med.* **2000**, *11*, 123–136. [CrossRef]
31. Cansiz, E.; Dogru, S.C.; Arslan, Y.Z. Evaluation of different fixation materials for mandibular condyle fractures. In Proceedings of the XIX International Conference on Mechanics in Medicine and Biology, Bologna, Italy, 3–5 September 2014.
32. Wirtz, D.C.; Schiffers, N.; Pandorf, T.; Radermacher, K.; Weichert, D.; Forst, R. Critical evaluation of known bone material properties to realize anisotropic FE-simulation of the proximal femur. *J. Biomech.* **2000**, *33*, 1325–1330. [CrossRef]
33. Schileo, E.; Dall'Ara, E.; Taddei, F.; Malandrino, A.; Schotkamp, T.; Baleani, M.; Viceconti, M. An accurate estimation of bone density improves the accuracy of subject-specific finite element models. *J. Biomech.* **2008**, *41*, 2483–2491. [CrossRef]
34. Helgason, B.; Perilli, E.; Schileo, E.; Taddei, F.; Brynjólfsson, S.; Viceconti, M. Mathematical relationships between bone density and mechanical properties: A literature review. *Clin. Biomech.* **2008**, *23*, 135–146. [CrossRef] [PubMed]
35. O'Mahony, A.M.; Williams, J.L.; Spencer, P. Anisotropic elasticity of cortical and cancellous bone in the posterior mandible increases peri-implant stress and strain under oblique loading. *Clin. Oral. Implant. Res.* **2001**, *12*, 648–657. [CrossRef] [PubMed]
36. Silva, G.C.; Cornacchia, T.M.; De Magalhaes, C.S.; Bueno, A.C.; Moreira, A.N. Biomechanical evaluation of screw- and cement-retained implant-supported prostheses: A nonlinear finite element analysis. *J. Prosthet. Dent.* **2014**, *112*, 1479–1488. [CrossRef] [PubMed]
37. Miura, J.; Maeda, Y.; Nakai, H.; Zako, M. Multiscale analysis of stress distribution in teeth under applied forces. *Dent. Mater.* **2009**, *25*, 67–73. [CrossRef]
38. Munari, L.S.; Cornacchia, T.P.; Moreira, A.N.; Gonçalves, J.B.; Casas, E.B.D.L.; Magalhães, C.S. Stress distribution in a premolar 3D model with anisotropic and isotropic enamel. *Med. Biol. Eng. Comput.* **2015**, *53*, 751–758. [CrossRef]

39. Su, K.C.; Chuang, S.F.; Ng, E.Y.K.; Chang, C.H. An investigation of dentinal fluid flow in dental pulp during food mastication: Simulation of fluid–structure interaction. *Biomech. Model. Mechanobiol.* **2014**, *13*, 527–535. [CrossRef] [PubMed]
40. Ha, S.-R.; Kim, S.-H.; Lee, J.-B.; Han, J.-S.; Yeo, I.-S.; Yoo, S.-H.; Kim, H.-K. Biomechanical three dimensional finite element analysis of monolithic zirconia crown with different cement thickness. *Ceram. Int.* **2016**, *42*, 14928–14936. [CrossRef]
41. Rocha, E.P.; Anchieta, R.B.; Almeida, E.O.; Freitas, A.C.F., Jr.; Martini, A.P.; Sotto-Maior, B.S.; Luersen, M.A.; Ko, C.C. Zirconia-based dental crown to support a removable partial denture: A three-dimensional finite element analysis using contact elements and micro-CT data. *Comput. Methods Biomech. Biomed. Eng.* **2015**, *18*, 1744–1752. [CrossRef]
42. Jithendra Babu, P.; Alla, R.K.; Alluri, V.R.; Datla, S.R.; Konakanchi, A. Dental ceramics: Part I—An overview of composition, structure and properties. *Am. J. Mater. Eng. Technol.* **2015**, *3*, 13–18.
43. Dejak, B.; Młotkowski, A.; Langot, C. Three-dimensional finite element analysis of molars with thin-walled prosthetic crowns made of various materials. *Dent. Mater.* **2012**, *28*, 433–441. [CrossRef] [PubMed]
44. Misch, C.E.; Qu, Z.; Bidez, M.W. Mechanical properties of trabecular bone in the human mandible: Implications for dental implant treatment planning and surgical placement. *J. Oral Maxillofac. Surg.* **1999**, *57*, 700–706; discussion 706–708. [CrossRef]
45. Morgan, E.F.; Unnikrisnan, G.U.; Hussein, A.I. Bone mechanical properties in healthy and diseased states. *Annu. Rev. Biomed. Eng.* **2018**, *20*, 119–143. [CrossRef]
46. Shimizu, K.; Oka, M.; Kumar, P.; Kotoura, Y.; Yamamuro, T.; Makinouchi, K.; Nakamura, T. Time-dependent changes in the mechanical properties of zirconia ceramic. *J. Biomed. Mater. Res.* **1993**, *27*, 729–734. [CrossRef] [PubMed]
47. Luthardt, R.G.; Holzhuter, M.; Sandkuhl, O.; Herold, V.; Schnapp, J.D.; Kuhlisch, E.; Walter, M. Reliability and properties of ground Y-TZP-zirconia ceramics. *J. Dent. Res.* **2002**, *81*, 487–491. [CrossRef]
48. Blatz, M.B.; Sadan, A.; Martin, J.; Lang, B. In vitro evaluation of shear bond strengths of resin to densely sintered high-purity zirconium-oxide ceramic after long-term storage and thermal cycling. *J. Prosthet. Dent.* **2004**, *91*, 356–362. [CrossRef] [PubMed]
49. Omur, C.; Gozneli, T.R.; Ozkan, Y. Effects of silica coating by physical vapor deposition and repeated firing on the low temperature degradation and flexural strength of a zirconia ceramic. *J. Prosthodont.* **2019**, *28*, e186–e194.
50. Goff, J.P.; Hayes, W.; Hull, S.; Hutchings, M.T.; Clausen, K.N. Defect structure of yttria stabilized zirconia and its influence on the ionic conductivity at elevated temperatures. *Phys. Rev. B* **1999**, *59*, 14202–14218. [CrossRef]
51. Alsadon, O.; Patrick, D.; Johnson, A.; Pollington, S.; Wood, D. Fracture resistance of zirconia-composite veneered crowns in comparison with zirconia-porcelain crowns. *Dent. Mater. J.* **2017**, *36*, 289–295. [CrossRef]
52. Sailer, B.; Pjetursson, E.; Zwahlen, Z.; Hammerle, C.H. A systematic review of the survival and complication rates of all ceramic and metal–ceramic reconstructions after an observation period of at least 3 years. Part II: Fixed dental prostheses. *Clin. Oral Implant. Res.* **2007**, *18*, 86–96. [CrossRef]
53. Larsson, C.; Von Steyern, P.V. Five-year follow-up of implant-supported Y-TZP and ZTA fixed dental prostheses. A randomized, prospective clinical trial comparing two different material systems. *Int. J. Prosthodont.* **2010**, *23*, 555–561.
54. Schwarz, S.; Schröder, C.; Hassel, A.; Bömicke, W.; Rammelsberg, P. Survival and chipping of zirconia based and metal–ceramic implant supported single crowns. *Clin. Implant. Dent. Relat. Res.* **2012**, *14*, e119–e125. [CrossRef]
55. Moraguez, O.D.; Wiskott, H.A.; Scherrer, S.S. Three-to nine-year survival estimates and fracture mechanisms of zirconia-and alumina-based restorations using standardized criteria to distinguish the severity of ceramic fractures. *Clin. Oral Investig.* **2015**, *19*, 2295–2307. [CrossRef]
56. Güncü, M.B.; Cakan, U.; Muhtarogullari, M.; Canay, S. Zirconia-based crowns up to 5 years in function: A retrospective clinical study and evaluation of prosthetic restorations and failures. *Int. J. Prosthodont.* **2015**, *28*, 152–157. [CrossRef]
57. Lin, M.T.; Sy-Munoz, J.; Munoz, C.A.; Goodacre, C.J.; Naylor, W.P. The effect of tooth preparation form on the fit of Procera copings. *Int. J. Prosthodont.* **1998**, *11*, 580–590.
58. Reich, S.; Wichmann, M.; Nkenke, E.; Proeschel, P. Clinical fit of all-ceramic three-unit fixed partial dentures, generated with three different CAD/CAM systems. *Eur. J. Oral Sci.* **2005**, *113*, 174–179. [CrossRef] [PubMed]
59. Pilo, R.; Cardash, H.S. In vivo retrospective study of cement thickness under crowns. *J. Prosthet. Dent.* **1998**, *79*, 621–625. [CrossRef] [PubMed]
60. Aboushelib, M.N.; De Jager, N.; Kleverlaan, C.J.; Feilzer, A.J. Effect of loading method on the fracture mechanics of two layered all-ceramic restorative systems. *Dent. Mater.* **2007**, *23*, 952–959. [CrossRef] [PubMed]
61. Casson, A.M.; Glyn Jones, J.C.; Youngson, C.C.; Wood, D.J. The effect of luting media on the fracture resistance of a flame sprayed all-ceramic crown. *J. Dent.* **2001**, *29*, 539–544. [CrossRef] [PubMed]
62. Sorrentino, R.; Triulzio, C.; Tricarico, M.G.; Bonadeo, G.; Gherlone, E.F.; Ferrari, M. In vitro analysis of the fracture resistance of CAD-CAM monolithic zirconia molar crowns with different occlusal thickness. *J. Mech. Behav. Biomed. Mater.* **2016**, *61*, 328–333. [CrossRef]
63. Magne, P. Efficient 3D finite element analysis of dental restorative procedures using micro-CT data. *Dent. Mater.* **2007**, *23*, 539–548. [CrossRef] [PubMed]
64. Himmlova, L.; Dostalova, T.; Kacovsky, A.; Konvickova, S. Influence of implant length and diameter on stress distribution: A finite element analysis. *J. Prosthet. Dent.* **2004**, *91*, 20–25. [CrossRef] [PubMed]
65. Geng, J.P.; Tan, K.B.; Liu, G.R. Application of finite element analysis in implant dentistry: A review of the literature. *J. Prosthet. Dent.* **2001**, *85*, 585–598. [CrossRef] [PubMed]

66. Heydecke, G.; Butz, F.; Binder, J.; Strub, J. Material characteristics of a novel shrinkage-free ZrSiO$_4$ ceramic for the fabrication of posterior crowns. *Dent. Mater.* **2007**, *23*, 785–791. [CrossRef]
67. Fathy, S.M.; Anwar, M.I.E.; Fallal, A.A.E.; El-Negoly, S.A. Three-dimensional finite element analysis of lower molar tooth restored with fully milled and layered zirconia crowns. *J. Dent. Health Oral Disord. Ther.* **2014**, *1*, 89–95. [CrossRef]
68. Rocha, E.P.; Anchieta, R.B.; Freitas, A.C., Jr.; de Almeida, E.O.; Cattaneo, P.M.; Ko, C.C. Mechanical behavior of ceramic veneer in zirconia-based restorations: A 3-dimensional finite element analysis using microcomputed tomography data. *J. Prosthet. Dent.* **2011**, *105*, 14–20. [CrossRef]
69. Taskonak, B.; Yan, J.; Mecholsky, J.J., Jr.; Sertqoz, A.; Koçak, A. Fractographic analyses of zirconia-based fixed partial dentures. *Dent. Mater.* **2008**, *24*, 1077–1082. [CrossRef]
70. De Jager, N.; Pallav, P.; Feilzer, A.J. The apparent increase of the Young's modulus in thin cement layers. *Dent. Mater.* **2004**, *20*, 457–462. [CrossRef]
71. Son, Y.H.; Han, C.H.; Kim, S. Influence of internal-gap width and cement type on the retentive force of zirconia copings in pullout testing. *J. Dent.* **2012**, *40*, 866–872. [CrossRef] [PubMed]
72. Toms, S.R.; Eberhardt, A.W. A nonlinear finite element analysis of the periodontal ligament under orthodontic tooth loading. *Am. J. Orthod. Dentofac. Orthop.* **2003**, *123*, 657–665. [CrossRef] [PubMed]
73. Ovy, E.G.; Romanyk, D.L.; Flores Mir, C.; Westover, L. Modelling and evaluating periodontal ligament mechanical behaviour and properties: A scoping review of current approaches and limitations. *Orthod. Craniofac. Res.* **2022**, *25*, 199–211. [CrossRef]
74. Karimi, A.; Razaghi, R.; Biglari, H.; Rahmati, S.M.; Sandbothe, A.; Hasani, M. Finite element modeling of the periodontal ligament under a realistic kinetic loading of the jaw system. *Saudi Dent. J.* **2020**, *32*, 349–356. [CrossRef] [PubMed]
75. Cattaneo, P.M.; Cornelis, M.A. Orthodontic tooth movement studied by finite element analysis: An update. What can we learn from these simulations? *Curr. Osteoporos. Rep.* **2021**, *19*, 175–181. [CrossRef] [PubMed]
76. Field, C.; Ichim, I.; Swain, M.V.; Chan, E.; Darendeliler, M.A.; Li, W.; Li, Q. Mechanical responses to orthodontic loading: A 3-dimensional finite element multi-tooth model. *Am. J. Orthod. Dentofac. Orthop.* **2009**, *135*, 174–181. [CrossRef]
77. Viecilli, R.F.; Katona, T.R.; Chen, J.; Hartsfield, J.K.; Roberts, W.E. Three-dimensional mechanical environment of orthodontic tooth movement and root resorption. *Am. J. Orthod. Dentofac. Orthop.* **2008**, *133*, 791.e11–791.e26. [CrossRef]

Disclaimer/Publisher's Note: The statements, opinions and data contained in all publications are solely those of the individual author(s) and contributor(s) and not of MDPI and/or the editor(s). MDPI and/or the editor(s) disclaim responsibility for any injury to people or property resulting from any ideas, methods, instructions or products referred to in the content.

Article

Simultaneous Osseo- and Odontointegration of Titanium Implants: Description of Two Cases in Human and Animal Models and Review of Their Experimental and Clinical Implications

Iván Valdivia-Gandur [1], María Cristina Manzanares-Céspedes [2,*], Wilson Astudillo-Rozas [1,2], Oscar Aceituno-Antezana [1,3], Victòria Tallón-Walton [2] and Víctor Beltrán [4,*]

1. Biomedical Department, Universidad de Antofagasta, Avenida Angamos 601, Antofagasta 1270300, Chile; ivan.valdivia@uantof.cl (I.V.-G.); wastudro11@alumnes.ub.edu (W.A.-R.); oscar.aceituno@uantof.cl (O.A.-A.)
2. Human Anatomy and Embryology Unit, Universitat de Barcelona, 08193 Barcelona, Spain; vtallon@ub.edu
3. Dentistry Department, Universidad de Antofagasta, Avenida Angamos 601, Antofagasta 1270300, Chile
4. Clinical Investigation and Dental Innovation Center (CIDIC), Dental School and Center for Translational Medicine (CEMT-BIOREN), Universidad de La Frontera, Temuco 4811230, Chile
* Correspondence: mcmanzanares@ub.edu (M.C.M.-C.); victor.beltran@ufrontera.cl (V.B.)

Abstract: Two cases of calcified bone and dental tissue integration with titanium implants are presented, along with a review of the literature on their experimental and clinical implications. First, histological analyses of a titanium implant extracted from a patient with iimplant disease revealed the integration of both dental and bone tissue on the implant's surface. Secondly, a biocompatibility study in an animal model documented two implants in contact with tooth roots. Samples from both animal and human models demonstrated simultaneous osseointegration and dental tissue neoformation, with the latter attributed to the activity of cementoblasts. The literature review confirms the formation of cementum around dental implants in contact with teeth. Certain clinical reports have proposed the insertion of implants into bone sites containing impacted teeth as a conservative treatment alternative, avoiding the need for tooth extraction surgery and demonstrating the successful integration of teeth, bone, and dental implants. Furthermore, the documented natural formation of periodontal tissues around dental implants provided a foundation for tissue engineering studies aimed at realizing implant–bone relationships similar to those of natural bone–tooth structures. The primary challenges remain the long-term preservation of periodontal-like tissue formed on implants and the imparting of functional proprioceptive properties.

Keywords: titanium implant; osseointegration; cementoconduction; cementointegration; dental tissue engineering; biomaterial

Citation: Valdivia-Gandur, I.; Manzanares-Céspedes, M.C.; Astudillo-Rozas, W.; Aceituno-Antezana, O.; Tallón-Walton, V.; Beltrán, V. Simultaneous Osseo- and Odontointegration of Titanium Implants: Description of Two Cases in Human and Animal Models and Review of Their Experimental and Clinical Implications. *Materials* 2024, 17, 5555. https://doi.org/10.3390/ma17225555

Academic Editors: Matej Par and Tobias Tauböck

Received: 16 October 2024
Revised: 6 November 2024
Accepted: 11 November 2024
Published: 14 November 2024

Copyright: © 2024 by the authors. Licensee MDPI, Basel, Switzerland. This article is an open access article distributed under the terms and conditions of the Creative Commons Attribution (CC BY) license (https://creativecommons.org/licenses/by/4.0/).

1. Introduction

The processes of osseointegration, osseoinduction, and osteoconduction have been extensively studied [1,2]. However, knowledge regarding the behavior of dental and periodontal tissues in contact with dental implants remains limited. Currently, at least three major points of discussion are associated with this phenomenon. The first is the incorporation of dental tissue within the mandibular or maxillary bone as part of the implant bed. The second is the application of tissue engineering to realize implant integration with bone tissue through periodontal-like tissue. The third is the preservation of a layer of dental tissue for alveolar ridge maintenance during implant bed preparation; this special layer is subsequently maintained in contact with the dental implant. These topics are of significant interest from histological, functional, and dental tissue engineering perspectives. The aim of this study is to report two cases of dental implant integration via dental, periodontal, and bone tissues in both a human and an animal model and to review the literature on the clinical and experimental implications of this phenomenon in the discussion.

2. Case Descriptions

2.1. Findings in Human Sample

A total of twelve implants diagnosed with peri-implantitis were extracted as part of a descriptive study on peri-implant diseases conducted at the Implant Clinic, Faculty of Dentistry, Universidad de La Frontera (UFRO), Temuco, Chile, and the Peri-implant Clinic, Faculty of Dentistry, Universidad de Concepción (UdeC), Concepción, Chile. The clinical protocol was approved by the Ethics Committee of Universidad de La Frontera (024-2018). The implants were extracted using a trephine of sufficient diameter to ensure a margin of at least 0.5 mm between the implant surface and the internal surface of the trephine. The samples were fixed in 4% buffered formalin for 72 h. For histological analysis, each complete specimen (implant plus surrounding tissue) was processed for plastic embedding. Subsequently, the samples were sectioned along the midline of the implant, dividing them into two segments. One segment was abraded, polished, and prepared for backscattered electron microscopy (BS-SEM) analysis. The other segment was used for histological analysis with trichrome staining, as described in the literature [3]. During analysis, one sample revealed that the implant made contact with the intraosseous remnant of a tooth root. (Figure 1). BS-SEM analysis revealed that bone tissue utilized the implant surface and dentin as scaffolds for regeneration (Figure 1A). Cementum was observed on the surfaces of the dentin and bone tissue (Figure 1C,D indicated by arrows). On the underside of the implant, an amorphous hard tissue resembling both bone and cementum occupied the space between the implant and dental tissue (Figure 1E,F).

2.2. Findings in the Animal Model

The biocompatibility study of dental implants with new surface treatments, conducted using two male pigs, was evaluated and approved by the Animal Experimentation Ethics Committee of the International University of Catalonia (06-2011). The animal protocols followed the principles of the 3Rs concerning the use of animal models in experimental research: Replacement, Reduction, and Refinement [4,5]. Furthermore, all procedures were carried out by specialized surgeons under veterinary supervision, and the care and management of the animals complied with the ethical standards outlined in the *Guide for the Care and Use of Laboratory Animals* [6]. A total of 24 implants (12 per animal) were placed in the maxillary and mandibular bones immediately following bilateral premolar extraction (Figure 2A). Postoperative lateral cephalic radiographs were taken to assess implant distribution (Figure 2B). Ninety days post-surgery, the animals were sacrificed, and the implants, along with surrounding tissues, were conditioned for embedding in light-curing resin for further analysis. The samples were then prepared for BS-SEM analysis and stained with toluidine blue for histological examination via light microscopy, as previously described [3].

Postoperative radiographs revealed that the implants placed anteriorly were situated near the canine teeth (implants 1.1, 2.1, 3.1, and 4.1 in Figure 2), which have long roots extending antero-posteriorly within both the maxillary and mandibular bones. Subsequently, during the sectioning process of samples embedded in plastic, implants 1.1 and 4.1 from specimen 1 were found to be in contact with dental tissues. Despite this, the animals exhibited normal behavior and feeding patterns, without any signs of discomfort during the observation period. Histological analysis revealed that implant 4.1 had penetrated the surface of the mandibular canine, while the right maxillary canine was superficially contacted by implant 1.1. In histological images obtained via BS-SEM (Figure 3A,B), tooth tissue formation was evident between the implant threads. Examinations of implant 4.1 and the surrounding tissue showed that bone, periodontal ligament, and tooth tissue (cementum) had adapted to the implant's surface (Figure 3C). In both implants, new cementum and bone were observed on their surfaces, originating from the adjacent tissue, indicating the simultaneous integration of the implant by both bone and tooth tissue. Furthermore, histological analyses of the mandibular canine revealed pulp tissue exhibiting characteristics of vitality (Figure 3D).

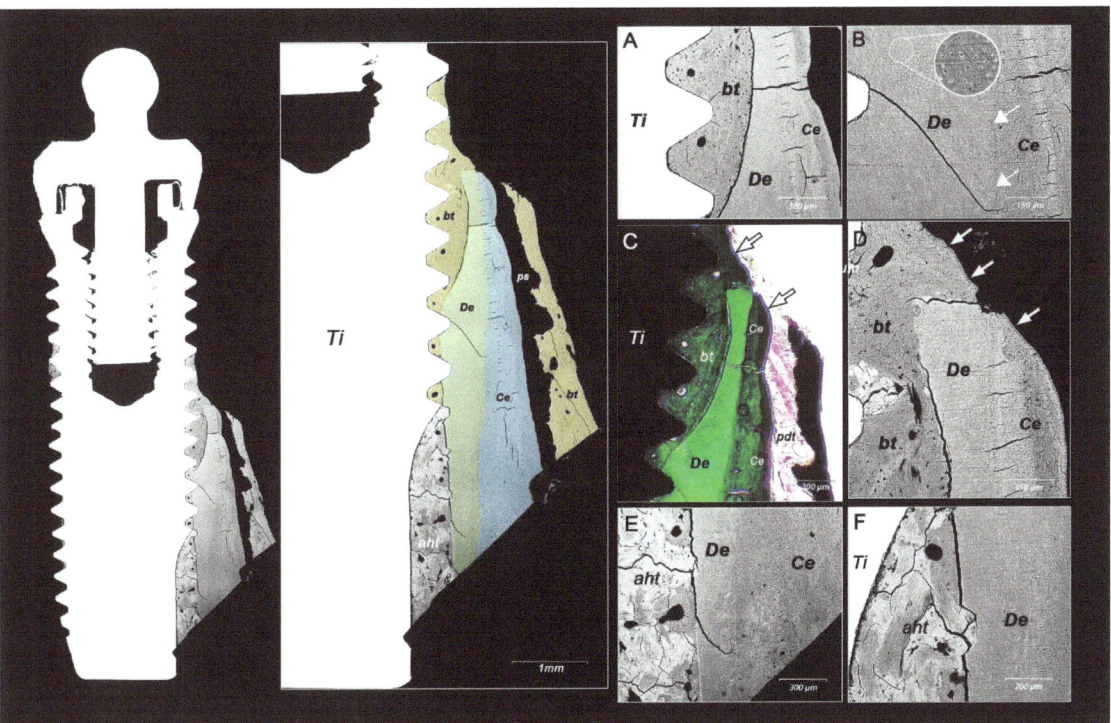

Figure 1. Dental tissue in contact with a dental implant in a human sample obtained after implant extraction due to peri-implant disease. The two images on the left display the complete sample, with dental tissue and bone in contact with the implant surface (Ti). Bone tissue (bt) is highlighted in yellow, dentin (De) in green, and cementum (Ce) in light blue. The periodontal space (ps) is visible near the dental tissue (cementum). The lower portion of the image shows an amorphous hard tissue (aht) containing bone tissue fragments, apparently integrated with newly formed cementum between the dentin and implant surface. Images (**A**–**F**) present histological details from the sample. In (**A**), bone tissue is observed between the implant threads and dental tissue (dentin and cementum). In (**B**), perforations on the surface indicate dentinal tubules (enhanced in the white circle with increased contrast); additionally, the boundary between dentin and cementum is marked (white arrows). In (**C**), trichrome staining reveals non-calcified periodontal tissue (pdt) compatible with the periodontal ligament, along with other hard tissues (cementum, dentin, and bone). In (**C**,**D**), cementum is observed over dentin and bone tissue (white arrows).

The BS-SEM images revealed differences in implant surface integration via osseous tissue (Figure 4A) and cementum (Figure 4B). The cementointegration process involved the use of dentin and cementum as a scaffold (Figure 3C) and direct integration with the implant's surface. The morphology of the cementum predominantly corresponded to that of acellular cementum. Furthermore, cementum demonstrated a greater capacity than bone for filling the spaces between the implant threads (Figures 3 and 4).

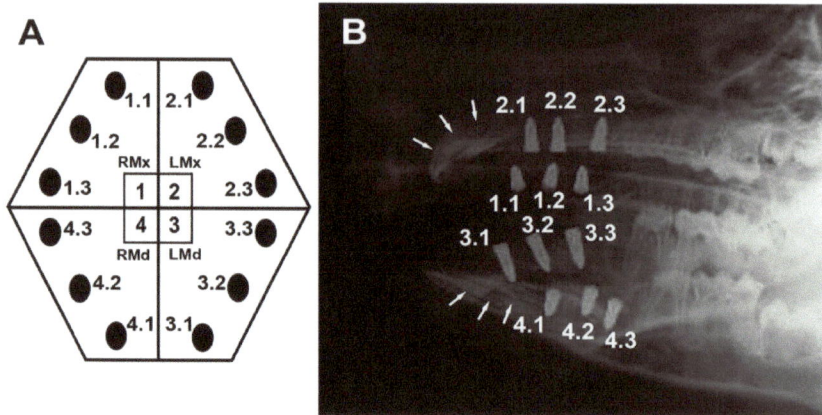

Figure 2. (**A**): Schematic illustration of implant distribution. Black circles indicate implant positions. RMx: Right maxilla; LMx: left maxilla; RMd: right mandible; LMd: left mandible. (**B**): Postoperative lateral radiograph showing implant distribution. White arrows indicate the position of the canines.

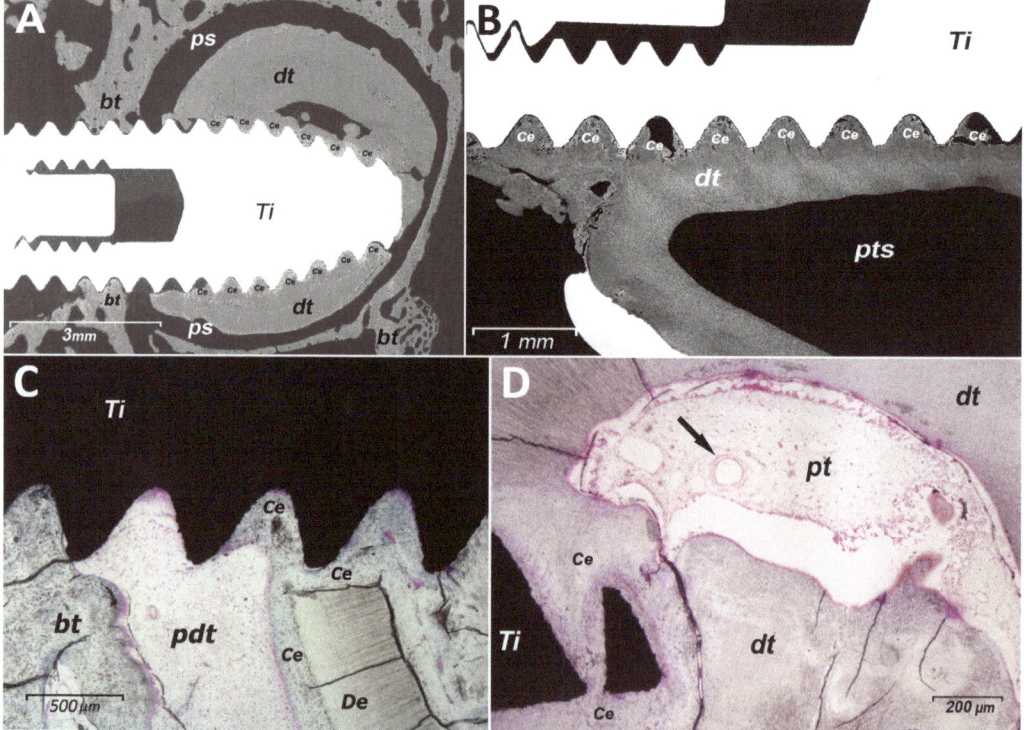

Figure 3. (**A**): SEM image of a titanium implant (Ti) 4.1 penetrating the canine tooth, surrounded by dental tissue (dt) and bone tissue (bt). (**B**): BS-SEM image showing dental tissue superficially eroded by implant 1.1, with its surface covered by cementum (Ce). (**C**): Histological image of implant 4.1 showing bone tissue and cementum formation on the surface of the implant threads (cementointegration), along with dentin (De). (**D**): Histological image of pulpal tissues (pt) in the periapical area of implant 4.1, stained with toluidine blue. The black arrow indicates a blood vessel. ps: Periodontal space; pdt: periodontal tissue; pts: pulpal tissue space.

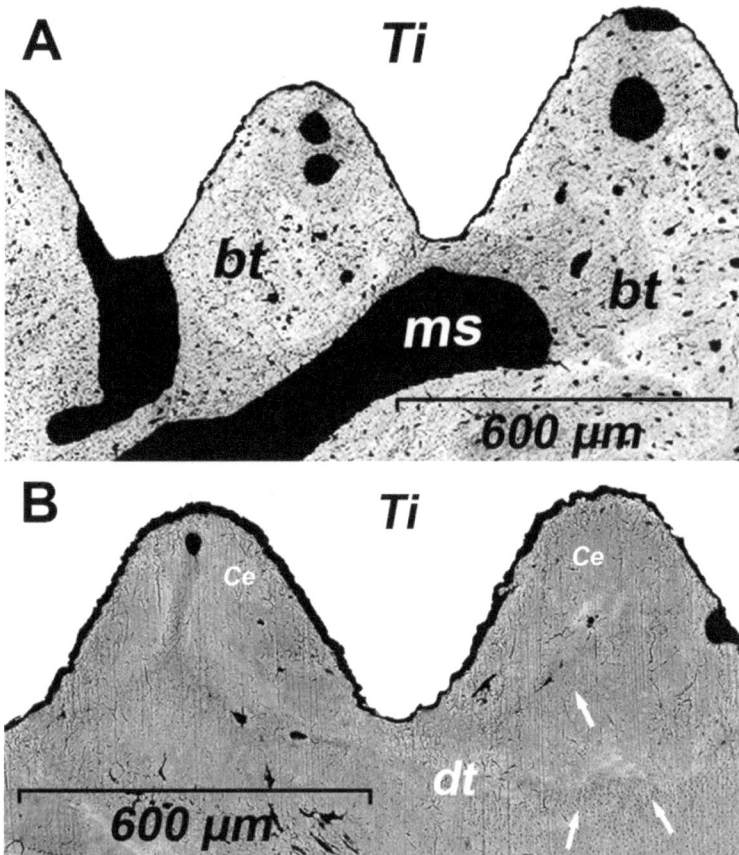

Figure 4. (**A**): BS-SEM image of threads from the osseointegrated dental implant 4.3 (specimen 1). (**B**): BS-SEM image illustrating dental tissue (dt) in contact with the implant's surface (implant 4.1), consistent with acellular cementum, demonstrating the cementointegration of the implant. Dentin is identified via the presence of dentinal tubules (white arrows). Cementum (Ce) is observed forming between the dentin and the implant surface. bt: Bone tissue; ms: marrow space; Ti: titanium dental implants.

3. Discussion

3.1. Experimental Implications on Dental and Periodontal Tissues Contacted by Dental Implants

In the animal findings described here, dental implants were integrated into periodontal tissues through the reparative activity of both soft tissue and cementoblasts following dental damage (Figures 2–4). This phenomenon demonstrates a pattern of cementum behavior similar to that of bone tissue on the implant surface, which allows it to be described as "cementointegration", a concept rarely addressed in the literature. Several studies have reported similar findings. Buser et al. [7,8] demonstrated the formation of periodontal tissue around dental implants, with cementum formation on cylindrical titanium implants in contact with retained dental roots, using a monkey model. Warrer et al. [9] obtained comparable results using a similar experimental design and animal model, but with a self-tapping screw-type implant system. In contrast, Gray and Vernino [10], in an analogous experiment with baboons, did not observe periodontal ligament formation, but did find cementum on the implant's surface. New cementum formation was also reported by Hürzeler et al. [11], who, using an animal model, inserted an implant into a dental socket with a retained root fragment in a procedure known as the "socket-shield technique", designed to prevent alveolar bone resorption. The formation of new cementum on both dental tissue (dentin) and implant surfaces is a consistent observation in experimental and incidental findings in the studies mentioned above.

Understanding how cementum initially contacts and continues to cover the implant surface and identifying which implant surface treatments or microenvironments best promote this phenomenon remain critical challenges. Urabe et al. [12] demonstrated that while the bioactivity of the implant material did not affect the migration of periodontium-derived cells, it significantly influenced cell differentiation. Hürzeler et al. [11] proposed a potential relationship between the use of enamel matrix derivatives and cementum formation on the implant surface.

The quantity of cementum may also be a critical factor. As illustrated in Figures 3 and 4, cementum demonstrated a greater capacity than bone to fill the space between the implant threads. These results are comparable to those observed in a study carried out by Hürzeler et al. [11].

The findings described here, along with the studies mentioned previously, demonstrate the response of vital cementum (characterized by the presence of active cementoblasts) in contact with dental implants. However, the effect of implants on non-vital cementum has not been as widely documented. In this context, the closest evidence was presented by Davarpanah et al. [13], who placed an implant in contact with an ankylosed dental root, achieving stability and realizing the integration of the implant with both dental and bone tissues. Similarly to the use of dentin as a biomaterial for the osteoconduction of new bone [14,15], the acellular cementum present in the implant bed may serve as a natural scaffold for bone regeneration

Although the characteristics of the implant's surface (such as material, porosity, and treatment) can influence both osseoconductivity and cementoconductivity, these phenomena also depend on specific molecular and cellular interactions. In a study by Parlar et al. [16], which investigated the formation of periodontal tissue on titanium surfaces, it was observed that the presence of periodontal tissue inhibited osseointegration. In this context, it is known that cementoblasts from the periodontal ligament express cementin 1 protein (CEMP1), which reduces the expression of Runx2 and osteocalcin genes [17], both of which are essential for osteoblast differentiation. Our findings in the animal model demonstrated that bone, periodontal ligament, and cementum healed and covered the implant surface, maintaining their respective positions even when the implant traversed the periodontal and dental structures (Figure 3C). In the human case, bone formation was observed utilizing both the implant surface and dentin as scaffolds, with cementum forming on dentin and bone tissue, although no direct contact was observed between cementum and the implant (Figure 4). Despite this, the amorphous hard tissue observed adjacent to the implant surface (aht, Figure 1) may represent tissue formation resulting from the complex interaction between bone and cementum. As a precedent, Rinaldi and Arana-Chavez [18] and Hürzeler et al. [11] described a dense amorphous material containing collagen fibrils, which formed a cementum-like layer over an implant in contact with the dental root. Further studies on this amorphous tissue are likely necessary to clarify the bone–cementum interaction during the integration of dental implants with bone and periodontal tissue. On the other hand, in experimental implant insertions that caused superficial damage to healthy tooth roots (as shown in Figure 3B), bone and periodontal tissue regeneration was observed, with a cementum layer formed on the implant surface but without significant alterations in tissue positions; both the tooth and implant remained healthy throughout the experimental period [18,19]. Similar results were obtained in experiments without traumatic contact between periodontal tissues and implant surfaces. Jahangiri et al. [20] demonstrated that implant–tooth contact was achieved by applying controlled orthodontic force to move the tooth toward the implant-induced cementum growth on the implant's surface, apparently through the transfer of cellular elements from the periodontal ligament.

Another experimental implication of the observations regarding implant–tooth contact relates to the formation of periodontal tissues on dental implants through a combination of in vitro and in vivo experimentation. For example, Choi [21] demonstrated that cultured periodontal ligament cells could induce the formation of cementum and ligament when placed on the implant surface and subsequently inserted into the mandibular bone of dogs. In another example, Marei et al. [22] placed implants combined with polymer scaffolds enriched with undifferentiated mesenchymal stem cells into the mandibular bone of goats, demonstrating that these cells were able to differentiate and promote the formation of cementum, bone, and periodontal ligament.

3.2. Clinical Implications of Contact Between Dental Tissues and Implants

There has been a growing interest in clinical approaches involving implant–tooth contact for various objectives (Table 1). Hürzeler et al. [11], followed by Bäumer et al. [23], implemented the socket-shield technique, which involves retaining a portion of the root in areas where the alveolar bone is thin during implant bed preparation, with the goal of preventing bone resorption. This method was initially evaluated in animal models, where it was reported that cementum-like tissue formed on the implant surface and that the gingival architecture surrounding the implant remained well preserved after six months. Additionally, there is evidence supporting an innovative approach for implant placement in bone sites with impacted (unerupted) teeth, where extraction is avoided to reduce the risk of bone loss (Table 1). However, a series of six cases (a total of eight implants) presented by Langer et al. [24] documented the failure of three implants inserted in anatomical regions with retained root remnants. Similarly, Guarnieri et al. [25] described the failure of a dental implant placed in contact with a dental root due to the development of peri-implant disease one year after insertion. Furthermore, evidence suggests that the devitalization of teeth caused by an implant passing through the root of erupted healthy teeth ultimately led to implant extraction [26–29]. Consequently, clinicians must consider varying tissue responses when an implant contacts dental tissue from impacted or unerupted teeth, erupted teeth, or retained root remnants.

Managing trauma and inflammation may optimize cementum growth on the implant surface in contact with dental tissue. The literature suggests that a proinflammatory stimulus (such as cytokines combined with compressive forces) can reduce the expression of bone sialoprotein and CEMP1 [30], both of which are essential for cementum formation [17,31]. Bone sialoprotein is a component of the extracellular matrix in mineralized tissues, and it plays a critical role in hydroxyapatite precipitation [32]. This protein is found in both cellular and acellular cementum structures [33], and its absence can result in significant defects in acellular cementum formation and periodontal attachment [31]. Moreover, Wang et al. [34] demonstrated through in vitro and in vivo studies that proinflammatory cytokines can impair cementum regeneration. Optimizing the conditions for cementum formation may even encourage more favorable coverage of the implant surface by cementum compared to bone. Studies on graft materials used for periodontal apparatus regeneration have shown that cementum covers the tooth surface faster and earlier than bone tissue [35,36]. This finding aligns with the observations in Figure 3, where cementum is seen to cover the implant surface more extensively than bone.

The manner in which the implant contacts the tooth and the status of the tooth in the oral cavity (unerupted, erupted, root remnant, etc.) appear to be important factors to consider in analyzing this subject. Table 1 schematically presents findings from the literature regarding the outcomes of titanium implants in contact with dental tissue in both animal and human models, taking into account the previously described variables.

Table 1. Literature evidence from experimental and clinical reports on titanium implants in contact with dental tissue (including our findings).

Dental Implant Contacting Dental or Periodontal Tissues	Evidence in the Literature Author (Year)	Human/Animal Evidence	Clinical Observations, Tissues Reactions, and Histological Evidence
Implant passing through the dental or periodontal tissue from root of erupted, functional teeth.	Sussman (1998 a y b) [26,27]	Human	One case: Implant passing through the root of an erupted mature tooth, causing a periapical lesion. Endodontic treatment and implant extraction were indicated. No histological evidence is available.
	Margelos and Verdelis (1995) [29]	Human	Three cases: Implant apparently passing through the periodontal tissue in the apical area caused irreversible pulpal damage. Endodontic treatment and implant extraction were indicated. No histological evidence is available.
	Our findings, Figures 2A,C,D and 4B	Animal	One implant passing through the root of an erupted tooth was integrated with both dental (cementum) and bone tissue.
Implant placed in bone sites with impacted teeth or supernumerary (passing through unerupted teeth)	Ouni and Mansour (2023) [37]	Human	One case: Two implants were placed through retained teeth in the mandibular bone of a patient with amelogenesis imperfecta. One implant evolved successfully (clinically stable after 36 months), while the other required replacement due to failure. No histological evidence is available.
	Brinkmann et al. (2020) [38]	Human	One case: Two implants were successfully placed through unerupted teeth in a patient with multiple impacted teeth. The implants remained clinically stable after 24 months. No histological evidence is available.
	Davarpanah et al. (2012, 2015) [13,39]	Human	Ten cases: A total of 15 implants were successfully placed through unerupted teeth. All implants remained clinically stable, with follow-up periods ranging from 1 to 8 years. No histological evidence is available (some cases were previously reported)
	Szmukler-Moncler et al. (2014) [40]	Human	One case: An implant was successfully placed through an unerupted tooth. The implant remained clinically stable after 18 months. No histological evidence is available.
	Kaplansky and Kurtzman (2024) [41]	Human	One case: An implant was placed in the maxillary anterior region, passing through supernumerary teeth. The implant remained clinically stable after 3 years and 8 months. No histological evidence is available.

Table 1. Cont.

Dental Implant Contacting Dental or Periodontal Tissues	Evidence in the Literature Author (Year)	Human/Animal Evidence	Clinical Observations, Tissues Reactions, and Histological Evidence
Implant passing through a dental root remnant (or retained root)	Our finding, Figure 1	Human	One case: Implant passing through a retained root. Implant was extracted due to peri-implant disease. Histological analysis showed new bone tissue and amorphous hard tissue (likely a mixture of cementum and bone tissue) in contact with the implant surface.
	Langer et al. (2015) [24]	Human	Six cases: Eight implants in contact with undetected root fragments. Osseointegration issues were observed in all implants, but only three were extracted. The remaining implants were surgically treated to remove the root remnant. Histological analysis of one sample showed acellular cementum on root fragments, with no histological evidence regarding the implant.
	Szmukler-Moncler et al. (2015) [42]	Human	Six cases: A total of seven implants were successfully placed through root remnants. All implants remained clinically stable, with follow-up ranging from 20 months to 9 years. No histological evidence is available.
	Baümer et al. (2013) [23]	Human	One case: Implant successfully placed using the socket-shield technique. Implant remained clinically stable after 6 months. No histological evidence is available.
	Davarpanah et al. (2012) [13]	Human	Two cases: Implants successfully placed through dental roots. One case involved an ankylosed root, and the other an endodontically treated root. Both implants remained clinically stable (after 32 and 20 months, respectively). No histological evidence is available.
	Hürzeler et al. (2010) [11]	Human	One case: Implant placed successfully using the socket-shield technique. Implant remained clinically stable after 6 months. No histological evidence is available.
	Davarpanah et al. (2009) [43]	Human	Five cases: Implants successfully placed through ankylosed dental roots. All implants remained clinically stable (follow-up ranged from 12 to 42 months). No histological evidence is available.
	Guarnieri et al. (2002) [25]	Human	One case: Implant was removed after one year due to peri-implant disease. Histological analysis showed formation of cellular cementum on the implant surface.

Table 1. Cont.

Dental Implant Contacting Dental or Periodontal Tissues	Evidence in the Literature Author (Year)	Human/Animal Evidence	Clinical Observations, Tissues Reactions, and Histological Evidence
Implant passing through a dental root remnant (or retained root)	Baümer et al. (2013) [23]	Animal	Twelve implants placed successfully using the socket-shield technique. Healthy periodontal tissues and new bone observed between implant and dentin. Cementum was not observed.
	Hürzeler et al. (2010) [11]	Animal	One implant successfully placed with the socket-shield technique, showing integration in dental and bone tissue. Areas between implant threads near the root fragment were partially filled with cellular cementum, amorphous mineralized tissue, and connective tissue.
	Gray and Vernino (2004) [10]	Animal	Ten implants successfully placed through remnant root tips. Implant integration in dental and bone tissue with cementoconduction on the implant surface was observed.
	Warrer et al. (1993) [9]	Animal	Eight implants successfully placed through remnant roots. Implant integration in dental and bone tissue with cementoconduction on the implant surface was observed. Also, formation of periodontal ligament was described.
	Buser et al. (1990a) [7]	Animal	Six implants successfully placed through remnant roots. Implant integration in dental and bone tissue with cementoconduction on the implant surface was observed. Also, formation of periodontal ligament was described.
Implant traumatically contacts the root surface of the tooth.	Our finding, Figure 3B	Animal	One implant contacting the root surface from erupted teeth was integrated by dental (cementum) and bone tissue.
	Urabe (2000) [12]	Animal	Twelve implants placed in contact with dental and periodontal tissue experimentally. Cementum-like tissue and periodontal ligament were observed mainly on the implant surface covered with hydroxyapatite.
	Asscherickx K et al. (2005) [19]	Animal	Three roots contacted by mini-implants for orthodontics, then removed. Histological examination of these roots demonstrated almost complete repair of the periodontal structure
	Rinaldi and Arana-Chávez (2010) [18]	Animal	Twenty-four mini-implants placed in contact with dental and periodontal tissue experimentally. The alveolar bone and periodontal ligament reorganized around the implant, forming a thin cementum-like layer over time at contact points with the periodontal ligament.

3.3. Applications and Challenges Arising from Evidence of Tooth–Implant Contact

Titanium dental implants achieve direct integration with bone but lack certain essential functional structures found in natural teeth, such as cementum and the periodontal ligament. These structures collectively function to absorb and buffer masticatory forces, which could provide the implant with properties that manage chewing stress, similarly to that endured by a natural tooth. The literature has introduced concepts such as the "functional periodontal ligament tissue formation on titanium implants" [44], the "biohybrid implant" [45,46] and the "functional implant" [47], developed through dental tissue engineering, which describe an implant that interacts with the bone through newly formed periodontal-like tissue. Washio et al. [45] demonstrated the creation of periodontal-like tissue around titanium implants under certain conditions using the "Cell Sheet Engineering Technology", transforming in tissue engineering what had been observed in different findings and clinical experiences (Table 1). Another concept associated with the creation of a peri-implant biological apparatus similar to dental periodontal tissue is that of "Ligaplant" [48,49]. Bio-hybrid implants and ligaplants, developed through tissue engineering, enable the construction of an implant-to-bone adhesion system similar to the natural tooth–bone union, incorporating cellular elements relevant for maintaining its physiology. However, achieving a more extensive coverage of the implant surface with cementum, similarly to that observed across the entire natural dental root, remains a considerable challenge. To date, it has been demonstrated that cementum has a greater capacity than bone to fill the space between the implant threads; nevertheless, it is mainly an acellular tissue and lacks the natural remodeling properties found in osseous tissue, for example. Therefore, at present, the coverage of implant surfaces with cementum is only relevant if it forms part of a more complex periodontal structure, including the periodontal ligament with cellular elements capable of participating in the remodeling and repair of the cementum formed on the implant surface [50]. Another significant challenge is realizing a sufficient functional proprioceptive response from the neo-structure to provide the implant with protection similar to that of a tooth with natural periodontal tissue.

Regarding advances in the science of biomaterials applied to new generations of dental implants, various options have been described that could contribute to the development of periodontal-like tissue around implants, such as the use of implant surface treatments via electrospinning, which enables the implant's surface to be loaded with biopolymers enriched with biomolecules [51] that could stimulate and protect tissue formation, or the use of nanoenzymes that promote tissue regeneration [52]. These surface treatments are independent of the material used to create the implant; however, it is important to consider that there is limited information on the behavior of periodontal tissue, particularly cementum, with respect to implants made from materials other than titanium.

3.4. Limitations

Although there is some clinical evidence regarding the viable placement of dental implants in contact with retained teeth or root remnants, establishing parameters for success or failure in this type of intervention is complex. The successful cases presented in the literature exhibit significant variations in terms of implant characteristics, type of rehabilitation applied to the implant, health status of the bone and dental structure where the implant was inserted, and patient history (age, general health, facial biotype, etc.). Consequently, the level of evidence is low, and the procedure cannot be standardized for consistent replication. Nevertheless, successful cases suggest that this treatment option could be considered a viable alternative subject to the patient's informed consent.

Regarding the experimental protocols analyzed, they generally converge on the concept of producing periodontal-like tissue around titanium implants. However, there is insufficient evidence to predict the long-term subsistence of the newly formed tissue under masticatory forces or its functional proprioceptive capacity. Moreover, it has been reported that these therapeutic alternatives have high costs and low predictability [48]. Finally, the use of titanium dental implants within this novel framework of interaction with human tissue raises new questions regarding biosafety, given the concerns expressed by some authors about the potential toxicity of the metal [50].

4. Conclusions

The findings described here, along with the literature review, indicate that titanium dental implants possess cementoconductive capacity. This property can be modified by altering the morphological or bioactive characteristics of the surface. Furthermore, it has been demonstrated that dental implants can integrate with both bone and dental tissue simultaneously when placed in contact with both structures under controlled conditions. In this context, clinical applications represent an intriguing treatment alternative that includes preparing an implant bed that traverses both the bone and dental tissue of impacted teeth or retained roots. Clinicians must consider that long-term evidence regarding the durability of this treatment alternative is limited. Additionally, the potential for periodontal tissue formation around dental implants has served as a foundation for tissue engineering studies aimed at achieving implant insertion in bone via periodontal-like tissue. The experimental success of tissue engineering in forming periodontal-like tissue for dental implant insertions requires controlled studies to assess its long-term utility under physiological conditions.

Author Contributions: Conceptualization, V.B., I.V.-G. and M.C.M.-C.; methodology, I.V.-G., V.B., V.T.-W., W.A.-R. and O.A.-A.; formal analysis, V.B. and I.V.-G.; investigation, I.V.-G., V.B., W.A.-R. and O.A.-A.; resources, I.V.-G., V.B. and M.C.M.-C.; data curation, I.V.-G., V.B. and V.T.-W.; writing—original draft preparation, I.V.-G. and V.B.; writing—review and editing, I.V.-G., V.B. and M.C.M.-C.; supervision, I.V.-G., V.B. and M.C.M.-C.; project administration, I.V.-G. and V.B.; funding acquisition, I.V.-G. and V.B. All authors have read and agreed to the published version of the manuscript.

Funding: This work was carried out Iván Valdivia-Gandur during a visit to the Universitat de Barcelona and was supported by the MINEDUC-UA project, code ANT 22991, and the National Council for Science and Technology Chile (Conicyt), project REDI170658. The animal experiment results reported were part of a project funded by AVINENT, S.A. (Manresa, Spain).

Institutional Review Board Statement: This study was conducted in accordance with the Declaration of Helsinki and approved by the Scientific Ethics Committee of the Universidad de La Frontera, Chile (Report N°024_2018.), and the ethical committee for Animal Experimentation of Cataluña University (Report N°06-2011).

Informed Consent Statement: Informed consent was obtained from all subjects involved in the study.

Data Availability Statement: The original contributions presented in the study are included in the article, further inquiries can be directed to the corresponding author.

Acknowledgments: The authors would express their gratitude to the Implant Clinic of the School of Dentistry, Universidad de La Frontera, Temuco, Chile, for their collaboration. The expert assistance of the Electron Microscopy Group of the Serveis Científico-Tècnics of the University of Barcelona is gratefully acknowledged.

Conflicts of Interest: The authors declare no conflicts of interest. The funders had no role in the design of the study; in the collection, analyses, or interpretation of data; in the writing of the manuscript; or in the decision to publish the results.

References

1. Albrektsson, T.; Johansson, C. Osteoinduction, Osteoconduction and Osseointegration. *Eur. Spine J.* **2001**, *10* (Suppl. S2), S96–S101. [PubMed]
2. Liu, J.; Kerns, D.G. Mechanisms of Guided Bone Regeneration: A Review. *Open Dent. J.* **2014**, *8*, 56–65. [CrossRef] [PubMed]
3. Manzanares, M.C.; Franch, J.; Carvalho, P.; Belmonte, A.M.; Tusell, J.; Franch, B.; Fernandez, J.M.; Clèries, L.; Morenza, J.L. BS-SEM Evaluation of the Tissular Interactions between Cortical Bone and Calcium-Phosphate Covered Titanium Implants. *Bull. Group Int. Rech. Sci. Stomatol. Odontol.* **2001**, *43*, 100–108. [PubMed]
4. Russell, W.M.S.; Burch, R. *The Principies of Humane Experimental Technique*; Methuen: London, UK, 1959.
5. Tannenbaum, J.; Bennett, B.T. Russell and Burch's 3Rs then and now: The need for clarity in definition and purpose. *J. Am. Assoc. Lab. Anim. Sci.* **2015**, *54*, 120–132. [PubMed]

6. National Research Council (US) Commitee for the Update of the Guide for the Care and Use of Laboratory Animals. *Guide for the Care and Use of Laboratory Animals*, 8th ed.; National Academies Press: Washington, DC, USA, 2011.
7. Buser, D.; Warrer, K.; Karring, T. Formation of a Periodontal Ligament around Titanium Implants. *J. Periodontol.* **1990**, *61*, 597–601. [CrossRef]
8. Buser, D.; Warrer, K.; Karring, T.; Stich, H. Titanium Implants with a True Periodontal Ligament: An Alternative to Osseointegrated Implants? *Int. J. Oral Maxillofac. Implants* **1990**, *5*, 113–116.
9. Warrer, K.; Karring, T.; Gotfredsen, K. Periodontal Ligament Formation around Different Types of Dental Titanium Implants. I. The Self-Tapping Screw Type Implant System. *J. Periodontol.* **1993**, *64*, 29–34. [CrossRef]
10. Gray, J.L.; Vernino, A.R. The Interface between Retained Roots and Dental Implants: A Histologic Study in Baboons. *J. Periodontol.* **2004**, *75*, 1102–1106. [CrossRef]
11. Hürzeler, M.B.; Zuhr, O.; Schupbach, P.; Rebele, S.F.; Emmanouilidis, N.; Fickl, S. The Socket-Shield Technique: A Proof-of-Principle Report. *J. Clin. Periodontol.* **2010**, *37*, 855–862. [CrossRef]
12. Urabe, M.; Hosokawa, R.; Chiba, D.; Sato, Y.; Akagawa, Y. Morphogenetic Behavior of Periodontium on Inorganic Implant Materials: An Experimental Study of Canines. *J. Biomed. Mater. Res.* **2000**, *49*, 17–24. [CrossRef]
13. Davarpanah, M.; Szmukler-Moncler, S.; Davarpanah, K.; Rajzbaum, P.; de Corbière, S.; Capelle-Ouadah, N.; Demurashvili, G. Unconventional transradicular implant placement to avoid invasive surgeries: Toward a potential paradigm shift. *Rev. Stomatol. Chir. Maxillofac.* **2012**, *113*, 335–349. [CrossRef] [PubMed]
14. Andrade, C.; Camino, J.; Nally, M.; Quirynen, M.; Martínez, B.; Pinto, N. Combining Autologous Particulate Dentin, L-PRF, and Fibrinogen to Create a Matrix for Predictable Ridge Preservation: A Pilot Clinical Study. *Clin. Oral Investig.* **2020**, *24*, 1151–1160. [CrossRef] [PubMed]
15. Um, I.-W.; Lee, J.-K.; Kim, J.-Y.; Kim, Y.-M.; Bakhshalian, N.; Jeong, Y.K.; Ku, J.-K. Allogeneic Dentin Graft: A Review on Its Osteoinductivity and Antigenicity. *Materials* **2021**, *14*, 1713. [CrossRef] [PubMed]
16. Parlar, A.; Bosshardt, D.D.; Unsal, B.; Cetiner, D.; Haytaç, C.; Lang, N.P. New Formation of Periodontal Tissues around Titanium Implants in a Novel Dentin Chamber Model. *Clin. Oral Implants Res.* **2005**, *16*, 259–267. [CrossRef] [PubMed]
17. Komaki, M.; Iwasaki, K.; Arzate, H.; Narayanan, A.S.; Izumi, Y.; Morita, I. Cementum Protein 1 (CEMP1) Induces a Cementoblastic Phenotype and Reduces Osteoblastic Differentiation in Periodontal Ligament Cells. *J. Cell. Physiol.* **2012**, *227*, 649–657. [CrossRef]
18. Rinaldi, J.C.; Arana-Chavez, V.E. Ultrastructure of the Interface between Periodontal Tissues and Titanium Mini-Implants. *Angle Orthod.* **2010**, *80*, 459–465. [CrossRef]
19. Asscherickx, K.; Vannet, B.V.; Wehrbein, H.; Sabzevar, M.M. Root Repair after Injury from Mini-Screw. *Clin. Oral Implants Res.* **2005**, *16*, 575–578. [CrossRef]
20. Jahangiri, L.; Hessamfar, R.; Ricci, J.L. Partial Generation of Periodontal Ligament on Endosseous Dental Implants in Dogs. *Clin. Oral Implants Res.* **2005**, *16*, 396–401. [CrossRef]
21. Choi, B.H. Periodontal Ligament Formation around Titanium Implants Using Cultured Periodontal Ligament Cells: A Pilot Study. *Int. J. Oral Maxillofac. Implants* **2000**, *15*, 193–196.
22. Marei, M.K.; Saad, M.M.; El-Ashwah, A.M.; El-Backly, R.M.; Al-Khodary, M.A. Experimental Formation of Periodontal Structure around Titanium Implants Utilizing Bone Marrow Mesenchymal Stem Cells: A Pilot Study. *J. Oral Implantol.* **2009**, *35*, 106–129. [CrossRef]
23. Bäumer, D.; Zuhr, O.; Rebele, S.; Schneider, D.; Schupbach, P.; Hürzeler, M. The Socket-Shield Technique: First Histological, Clinical, and Volumetrical Observations after Separation of the Buccal Tooth Segment—A Pilot Study. *Clin. Implant Dent. Relat. Res.* **2015**, *17*, 71–82. [CrossRef] [PubMed]
24. Langer, L.; Langer, B.; Salem, D. Unintentional Root Fragment Retention in Proximity to Dental Implants: A Series of Six Human Case Reports. *Int. J. Periodontics Restor. Dent.* **2015**, *35*, 305–313. [CrossRef] [PubMed]
25. Guarnieri, R.; Giardino, L.; Crespi, R.; Romagnoli, R. Cementum Formation around a Titanium Implant: A Case Report. *Int. J. Oral Maxillofac. Implants* **2002**, *17*, 729–732. [PubMed]
26. Sussman, H.I. Tooth Devitalization via Implant Placement: A Case Report. *Periodontal Clin. Investig.* **1998**, *20*, 22–24. [PubMed]
27. Sussman, H.I. Periapical Implant Pathology. *J. Oral Implantol.* **1998**, *24*, 133–138. [CrossRef]
28. Kim, S.G. Implant-Related Damage to an Adjacent Tooth: A Case Report. *Implant Dent.* **2000**, *9*, 278–280. [CrossRef]
29. Margelos, J.T.; Verdelis, K.G. Irreversible Pulpal Damage of Teeth Adjacent to Recently Placed Osseointegrated Implants. *J. Endod.* **1995**, *21*, 479–482. [CrossRef]
30. Diercke, K.; König, A.; Kohl, A.; Lux, C.J.; Erber, R. Human Primary Cementoblasts Respond to Combined IL-1β Stimulation and Compression with an Impaired BSP and CEMP-1 Expression. *Eur. J. Cell Biol.* **2012**, *91*, 402–412. [CrossRef]
31. Foster, B.L.; Soenjaya, Y.; Nociti, F.H.; Holm, E.; Zerfas, P.M.; Wimer, H.F.; Holdsworth, D.W.; Aubin, J.E.; Hunter, G.K.; Goldberg, H.A.; et al. Deficiency in Acellular Cementum and Periodontal Attachment in Bsp Null Mice. *J. Dent. Res.* **2013**, *92*, 166–172. [CrossRef]
32. Ganss, B.; Kim, R.H.; Sodek, J. Bone Sialoprotein. *Crit. Rev. Oral Biol. Med.* **1999**, *10*, 79–98. [CrossRef]
33. Macneil, R.L.; Sheng, N.; Strayhorn, C.; Fisher, L.W.; Somerman, M.J. Bone Sialoprotein Is Localized to the Root Surface during Cementogenesis. *J. Bone Miner. Res.* **1994**, *9*, 1597–1606. [CrossRef] [PubMed]
34. Wang, Y.; Li, Y.; Shao, P.; Wang, L.; Bao, X.; Hu, M. IL1β inhibits differentiation of cementoblasts via microRNA-325-3p. *J. Cell. Biochem.* **2020**, *121*, 2606–2617. [CrossRef] [PubMed]

35. Artzi, Z.; Wasersprung, N.; Weinreb, M.; Steigmann, M.; Prasad, H.S.; Tsesis, I. Effect of Guided Tissue Regeneration on Newly Formed Bone and Cementum in Periapical Tissue Healing after Endodontic Surgery: An in Vivo Study in the Cat. *J. Endod.* **2012**, *38*, 163–169. [CrossRef] [PubMed]
36. von Arx, T.; Britain, S.; Cochran, D.L.; Schenk, R.K.; Nummikoski, P.; Buser, D. Healing of Periapical Lesions with Complete Loss of the Buccal Bone Plate: A Histologic Study in the Canine Mandible. *Int. J. Periodontics Restor. Dent.* **2003**, *23*, 157–167.
37. Ouni, I.; Mansour, L. Inappropriate Protocol of Implant Placement in Contact with Impacted Teeth Leading to Failure. *Case Rep. Dent.* **2023**, *20*, 7328891. [CrossRef]
38. Brinkmann, J.C.-B.; Lobato-Peña, M.; Pérez-González, F.; Molinero-Mourelle, P.; Sánchez-Labrador, L.; Santos-Marino, J.; López-Quiles, J.; Martínez-González, J.M. Placing Dental Implants through Impacted Teeth to Support a Fixed Partial Denture in a Geriatric Patient as an Alternative to Invasive Extraction Surgeries. *Eur. J. Dent.* **2020**, *14*, 697–701. [CrossRef]
39. Davarpanah, M.; Szmukler-Moncler, S.; Rajzbaum, P.; Davarpanah, K.; Capelle-Ouadah, N.; Demurashvili, G. Unconventional Implant Placement. V: Implant Placement through Impacted Teeth; Results from 10 Cases with an 8- to 1-Year Follow-Up. *Int. Orthod.* **2015**, *13*, 164–180. [CrossRef]
40. Szmukler-Moncler, S.; Davarpanah, M.; Davarpanah, K.; Rajzbaum, P.; Capelle-Ouadah, N.; Demurashvili, G. Implants in contact with tissues other than bone. Is there room for a potential paradigm shift? *Swiss Dent. J.* **2014**, *124*, 149–164.
41. Kaplansky, I.V.; Kurtzman, G.M. Implant Placement When an Impacted Tooth and Supernumerary Teeth Are Present in the Maxilla. *Compend. Contin. Educ. Dent.* **2024**, *45*, e1–e4.
42. Szmukler-Moncler, S.; Davarpanah, M.; Davarpanah, K.; Capelle-Ouadah, N.; Demurashvili, G.; Rajzbaum, P. Unconventional Implant Placement Part III: Implant Placement Encroaching upon Residual Roots—A Report of Six Cases. *Clin. Implant Dent. Relat. Res.* **2015**, *17* (Suppl. S2), e396–e405. [CrossRef]
43. Davarpanah, M.; Szmukler-Moncler, S. Unconventional implant treatment: I. Implant placement in contact with ankylosed root fragments. A series of five case reports. *Clin. Oral Implants Res.* **2009**, *20*, 851–856. [CrossRef] [PubMed]
44. Lin, Y.; Gallucci, G.O.; Buser, D.; Bosshardt, D.; Belser, U.C.; Yelick, P.C. Bioengineered Periodontal Tissue Formed on Titanium Dental Implants. *J. Dent. Res.* **2011**, *90*, 251–256. [CrossRef] [PubMed]
45. Washio, K.; Tsutsumi, Y.; Tsumanuma, Y.; Yano, K.; Srithanyarat, S.S.; Takagi, R.; Ichinose, S.; Meinzer, W.; Yamato, M.; Okano, T.; et al. In Vivo Periodontium Formation Around Titanium Implants Using Periodontal Ligament Cell Sheet. *Tissue Eng. Part A* **2018**, *24*, 1273–1282. [CrossRef] [PubMed]
46. Oshima, M.; Inoue, K.; Nakajima, K.; Tachikawa, T.; Yamazaki, H.; Isobe, T.; Sugawara, A.; Ogawa, M.; Tanaka, C.; Saito, M.; et al. Functional Tooth Restoration by Next-Generation Bio-Hybrid Implant as a Bio-Hybrid Artificial Organ Replacement Therapy. *Sci. Rep.* **2014**, *4*, 6044. [CrossRef] [PubMed]
47. Iwasaki, K.; Washio, K.; Meinzer, W.; Tsumanuma, Y.; Yano, K.; Ishikawa, I. Application of Cell-Sheet Engineering for New Formation of Cementum around Dental Implants. *Heliyon* **2019**, *5*, e01991. [CrossRef]
48. Saleem, M.; Kaushik, M.; Ghai, A.; Tomar, N.; Singh, S. Ligaplants: A Revolutionary Concept in Implant Dentistry. *Ann. Maxillofac. Surg.* **2020**, *10*, 195–197. [CrossRef]
49. Bajaj, P.; Shirbhate, U.; Dare, S. Ligaplants: Uprising Regimen in the Glebe of Implant Dentistry. *Cureus* **2023**, *15*, e45968. [CrossRef]
50. Seo, B.M.; Miura, M.; Gronthos, S.; Bartold, P.M.; Batouli, S.; Brahim, J.; Young, M.; Robey, P.G.; Wang, C.Y.; Shi, S. Investigation of multipotent postnatal stem cells from human periodontal ligament. *Lancet* **2004**, *364*, 149–155. [CrossRef]
51. Ozkan, A.; Çakır, D.A.; Tezel, H.; Sanajou, S.; Yirun, A.; Baydar, T.; Erkekoglu, P. Dental Implants and Implant Coatings: A Focus on Their Toxicity and Safety. *J. Environ. Pathol. Toxicol. Oncol.* **2023**, *42*, 31–48. [CrossRef]
52. Zhu, B.; Li, L.; Wang, B.; Miao, L.; Zhang, J.; Wu, J. Introducing Nanozymes: New Horizons in Periodontal and Dental Implant Care. *Chembiochem* **2023**, *3*, 24e202200636. [CrossRef]

Disclaimer/Publisher's Note: The statements, opinions and data contained in all publications are solely those of the individual author(s) and contributor(s) and not of MDPI and/or the editor(s). MDPI and/or the editor(s) disclaim responsibility for any injury to people or property resulting from any ideas, methods, instructions or products referred to in the content.

MDPI AG
Grosspeteranlage 5
4052 Basel
Switzerland
Tel.: +41 61 683 77 34

Materials Editorial Office
E-mail: materials@mdpi.com
www.mdpi.com/journal/materials

Disclaimer/Publisher's Note: The title and front matter of this reprint are at the discretion of the Guest Editors. The publisher is not responsible for their content or any associated concerns. The statements, opinions and data contained in all individual articles are solely those of the individual Editors and contributors and not of MDPI. MDPI disclaims responsibility for any injury to people or property resulting from any ideas, methods, instructions or products referred to in the content.

www.ingramcontent.com/pod-product-compliance
Lightning Source LLC
LaVergne TN
LVHW072358090526
838202LV00019B/2575